华西医学大系

解读"华西现象"

讲述华西故事

展示华西成果

向四川大学华西医院建院130周年献礼
For the 130th Anniversary of
West China Hospital, Sichuan University

华西坝老建筑的前世今生
THE STORIES OF THE HISTORIC ARCHITECTURE OF HUAXIBA

（汉英对照）
(Chinese-English Bilingual Edition)

戚亚男 / 著　罗俊荷 / 译

四川科学技术出版社
·成都·

图书在版编目（CIP）数据

华西坝老建筑的前世今生＝THE STORIES OF THE HISTORIC ARCHITECTURE OF HUAXIBA (Chinese-English Bilingual Edition)：汉英对照 / 戚亚男著；罗俊荷译. -- 成都：四川科学技术出版社，2022.10

（华西医学大系）

ISBN 978-7-5727-0717-9

Ⅰ.①华… Ⅱ.①戚…②罗… Ⅲ.①四川大学华西临床医学院—建筑物—介绍—民国—汉、英 Ⅳ.①TU246.1

中国版本图书馆CIP数据核字(2022)第174912号

成都市哲学社会科学重点研究基地成都历史与成都文献研究中心2021年度规划项目"华西坝老建筑里的历史人物"（项目编号：CLWX21006）

华西坝老建筑的前世今生（汉英对照）

THE STORIES OF THE HISTORIC ARCHITECTURE OF HUAXIBA
（Chinese-English Bilingual Edition）

HUAXIBA LAOJIANZHU DE QIANSHI JINSHENG（HAN-YING DUIZHAO）

戚亚男／著　罗俊荷／译

出 品 人	程佳月
策划组稿	钱丹凝　程佳月
责任编辑	王双叶
责任校对	王　川　张　琪
封面设计	经典记忆
版式设计	成都华桐美术设计有限公司
责任出版	欧晓春
出版发行	四川科学技术出版社
地　　址	四川省成都市锦江区三色路238号新华之星A座 传真：028-86361756　邮政编码：610023
成品尺寸	156mm×236mm
印　　张	19　字　数　450千
印　　刷	成都市金雅迪彩色印刷有限公司
版　　次	2022年10月第1版
印　　次	2023年9月第1次印刷
定　　价	128.00元

ISBN 978-7-5727-0717-9

■ 版权所有　翻印必究 ■

邮购：四川省成都市锦江区三色路238号新华之星A座25层
邮购电话：86361770　邮政编码：610023

《华西医学大系》顾问

(按姓氏笔画为序)

马俊之　吕重九　李　虹　步　宏　何志勇
张　伟　张肇达　陈钟光　郑尚维　胡富合
殷大奎　唐孝达　曹泽毅　敬　静　魏于全

《华西医学大系》编委会

主任委员

李正赤　李为民　罗　勇　刘龙章　李　强

副主任委员

罗凤鸣　刘伦旭　黄　进　陈大利　王华光　雷　华

委　员

程南生　申文武　胡建昆　王坤杰　吴　泓　郭应强
陈　蕾　余　淳　钱丹凝　秦伏男　程佳月

秘书组

韩　平　赵　欣　郑　源　卢添林　周　亮
李雷雷　唐绍军　杜　晨　刘　沁　张茜惠
罗小燕　何晓霞　李　栎　税萌成　兰　银

《华西医学大系》总序

由四川大学华西临床医学院/华西医院（简称"华西"）与新华文轩出版传媒股份有限公司（简称"新华文轩"）共同策划、精心打造的《华西医学大系》陆续与读者见面了，这是双方强强联合，共同助力健康中国战略、推动文化大繁荣的重要举措。

百年华西，历经120多年的历史与沉淀，华西人在每一个历史时期均辛勤耕耘，全力奉献。改革开放以来，华西励精图治、奋进创新，坚守"关怀、服务"的理念，遵循"厚德精业、求实创新"的院训，为践行中国特色卫生与健康发展道路，全心全意为人民健康服务做出了积极努力和应有贡献，华西也由此成为了全国一流、世界知名的医（学）院。如何继续传承百年华西文化，如何最大化发挥华西优质医疗资源辐射作用？这是处在新时代站位的华西需要积极思考和探索的问题。

新华文轩，作为我国首家"A+H"出版传媒企业、中国出版发行业排头兵，一直都以传承弘扬中华文明、引领产业发展为使命，以坚持导向、服务人民为己任。进入新时代后，新华文轩提出了坚持精准出版、精细出版、精品出版的"三精"出版发展思路，全心全意为推动我国文化发展与繁荣做出了积极努力和应有贡献。如何充分发挥新华文轩的出版和渠道优势，不断满足人民日益增长的美好生活需要？这是新华文轩一直以来积极思考和探索的问题。

基于上述思考，四川大学华西临床医学院/华西医院与新华文轩出版传媒股份有限公司于2018年4月18日共同签署了战略合作协议，启动了《华西医学大系》出版项目并将其作为双方战略合作的重要方面和旗舰项目，共同向承担《华西医学大系》出版工作的四川科学技术出版社授予了"华西医学出版中心"铭牌。

人民健康是民族昌盛和国家富强的重要标志，没有全民健康，就没有全面小康，医疗卫生服务直接关系人民身体健康。医学出版是医药卫生事业发展的重要组成部分，不断总结医学经验，向学界、社会推广医学成果，普及医学知识，对我国医疗水平的整体提高、对国民健康素养的整体提升均具有重要的推动作用。华西与新华文轩作为国内有影响力的大型医学健康机构与大型文化传媒企业，深入贯彻落实健康中国战略、文化强国战略，积极开展跨界合作，联合打造《华西医学大系》，展示了双方共同助力健康中国战略的开阔视野、务实精神和坚定信心。

华西之所以能够成就中国医学界的"华西现象"，既在于党政同心、齐抓共管，又在于华西始终注重临床、教学、科研、管理这四个方面协调发展、齐头并进。教学是基础，科研是动力，医疗是中心，管理是保障，四者有机结合，使华西人才辈出，临床医疗水平不断提高，科研水平不断提升，管理方法不断创新，核心竞争力不断增强。

《华西医学大系》将全面系统深入展示华西医院在学术研究、临床诊疗、人才建设、管理创新、科学普及、社会贡献等方面的发展成就；是华西医院长期积累的医学知识产权与保护的重大项目，是华西医院品牌建设、文化建设的重大项目，也是讲好"华西故事"、展示"华西人"风采、弘扬"华西精神"的重大项目。

《华西医学大系》主要包括以下子系列。

①《学术精品系列》：总结华西医（学）院取得的学术成果，学术影响力强。②《临床实用技术系列》：主要介绍临床各方面的适宜技术、新技术等，针对性、指导性强。③《医学科普系列》：聚焦百姓最关心的、最迫切需要的医学科普知识，以百姓喜闻乐见的方式呈现。④《医院管理创新系列》：展示华西医（学）院管理改革创新的系列成果，体现华西"厚德精业、求实创新"的院训，探索华西医院管理创新成果的产权保护，推广华西优秀的管理理念。⑤《精准医疗扶贫系列》：包括华西特色智力扶贫的相关内容，旨在提高贫困地区基层医院的临床诊疗水平。⑥《名医名家系列》：展示华西人的医学成就、贡献和风采，弘扬华西精神。⑦《百年华西系列》：聚焦百年华西历史，书写百年华西故事。

我们将以精益求精的精神和持之以恒的毅力精心打造《华西医学大系》，将华西的医学成果转化为出版成果，向西部、全国乃至海外传播，提升我国医疗资源均衡化水平，造福更多的患者，推动我国全民健康事业向更高的层次迈进。

《华西医学大系》编委会
2018年7月

序

  1913年2月1日，星期六，一位53岁的建筑师和他的兄弟乔治登上了从伦敦开往哈里奇港口的火车，开始了一场不可思议的冒险旅程。建筑师的名字叫荣杜易，他在中国成都的一所新大学的建筑设计竞标赛中获胜。这所学校就是由4个基督教差会创办的华西协合大学。

  现在从伦敦直飞成都用不到11个小时，然而在1913年，这段旅程无比艰难。①从哈里奇出发，荣杜易和他的兄弟乘船到荷兰角，然后乘火车经过柏林、华沙、莫斯科和西伯利亚地区到达北京。②他们于2月14日抵达北京，在北京会见了相关的人们，参观了今天仍有大批游客涌向的景点。2月25日，他们乘火车出发，于27日抵达汉口。3月1日，他们乘轮船逆长江而上到达宜昌，然后换乘小木船，由一队纤夫拖着通过长江三峡。他们乘坐的船曾经陷入一个危险的漩涡，一路艰险，终于在3月22日抵达万县。他们将从这里出发穿越乡村城镇去目的地成都。路途中他们有时坐着轿子，有时住在当地的旅馆或寺庙里。他们还在遂宁考察了当地的建筑风格和方法，并于4月8日抵达成都。

  这似乎很奇怪，经过2个多月的旅行，荣杜易和乔治在成都待了不到一个月就于5月3日离开了。在这短短的时间里，荣杜易研究了修建大学建筑的场地及其复杂的情况；出席了争论不休的会议，继而对当地人有所了解；与大学创办人员、当地政府官员和传教士对大学建筑的相关事宜进行了讨论。荣杜易还抽出时间为四川省民政长起草了少城公园（现人民公园）的建筑规划，并与民政长讨论了花园城市的概念（当时荣杜易在英国参与了一项建设花园城市的活动）；还参与了成都其他建筑的设计，例如春熙路附近的基督教青年会大楼。

注：此文是赵安祝、赵丽勤为本书所撰英文序的中文译文，译文对原序作了适当修改。

回程的时间更长。荣杜易和乔治乘小木船到宜昌，然后转乘轮船，在5月27日抵达上海。乔治直接乘船回家了，而荣杜易乘船经过日本、香港和檀香山（火奴鲁鲁），于7月9日到达旧金山，然后乘火车去多伦多，于7月21日到达纽约，与资助该项目的传教士团体进行会谈。他终于在1913年7月底回到了家中。

此后，荣杜易再也没有回到中国。但是他继续推动着这个项目，从他位于伦敦哈默史密斯街11号的办公室接收和发送了来往于英国与中国之间的无数信件、照片和图纸。他注意到建筑过程中的任何小事，并详细说明了建筑装饰的细节。他也面临着挑战，比如施工现场发现了古老的坟墓，就必须要修改修建计划，并与当地社区进行协商。难以置信的是，这样一个庞大而复杂的项目竟然可以通过信件远程指导。荣杜易一直在负责指导华西协合大学的建筑工作，直到1927年去世，他的儿子道格拉斯·伍德维尔·朗特里接管了他的工作。

虽然荣杜易是这个建筑项目的主要负责人，但这个项目团队的其他人也很重要。与他一起工作的伦敦的同事中最重要的是他的儿子道格拉斯，以及画了很多建筑图纸的弗兰克·奥斯勒。而把他的设计变成现实的成都当地人则更为重要。这就包括为了建设这个学校项目的当地管理人员和数百名建筑工人。此外，其他建筑师也对校园建设做出了贡献，书中描述的建筑中有4座不是由荣杜易设计的。

荣杜易何许人也？[③]早在1913年，荣杜易就是一名声名显赫的建筑师了。他的具有工艺美术风格的作品在英国很受欢迎。工艺美术运动以威廉·莫里斯等人的思想和作品为基础，强调手工制作、精心制作和物品的重要性，而不是工厂大批量生产。在建筑方面，它鼓励使用传统的建筑技术，简单的设计，使用当地材料，并拒绝任何强加的建筑风格。除了对工业化影响的担忧外，工艺美术运动还有政治方面的因素，其追随者的动机是希望工业和当地居民的生活环境都人性化。

在荣杜易为华西协合大学设计的作品中，我们可以看到许多这种工艺美术的传统。他的作品曾遭到大学理事会的批评，他们想要的是"现代西方"的建筑，而荣杜易坚持工艺美术原则。他在中国旅行途中花时间研究成都当地的传统、材料和建筑风格，从他在旅行途中买回来的物品中，我们知道他还学习了中国传统建筑。本书描述的荣杜易所设计的华西协合大学的建筑，正是他把英国工艺美术的理念与中国传统建筑设计融合的产物。

荣杜易出生在英国北约克郡海边斯卡伯勒小镇的一个杂货商家庭。他的大家庭包括居住在约克郡的荣杜易家族，他们是英国著名的巧克力制造商，也是慈善家和社会改革家。荣杜易全家都是贵格会信徒。

荣杜易的宗教信仰深深地影响了他的工作，他的大部分工作是由他的社会责任感推动的。他是建立约旦村的主要推动者，这是一个由当地居民建造和经营的模范村庄。他为第一次世界大战中的难民设计了预制房屋，并帮助建立了一个受伤士兵康复中心。他最大的兴趣之一是汉普郡之家，这是一个为来自伦敦最贫困地区的人们提供娱乐、教育和研讨会的俱乐部。1913年，他从中国寄回伦敦的许多信件都是写给与汉普郡之家有关的人。在中国，他同样受到一种激励，那就是给人们提供在大学接受教育的机会。

人们对教育与大学的需求随着时间的推移而有所改变。华西协合大学不断发展，现在是一所著名的医学院，即四川大学的一部分。这本书展示了华西协合大学的建筑是如何以不同的方式被使用的，以及在这些建筑中学习和工作的人们的一些鼓舞人心的故事。

荣杜易是赵丽勤的曾祖父和赵安祝的高祖父。赵安祝有幸两次访问四川大学，作为一名学者，他很高兴地看到在四川大学进行的各种研究，以及在华西医院进行的医疗工作。而赵丽勤只是短暂地访问了华西校园。我们相信，荣杜易会很高兴看到他设计的建筑仍然具有使用价值，人们逐渐会欣赏它们的美丽和价值。

赵安祝第二次访问四川大学时，和他的妻子凯瑟琳·厄奇教授向学校赠送了一系列与华西协合大学的老建筑有关的平面图、照片、图纸和文件，这些都是赵丽勤多年来保管和珍藏的。在这本书中，戚亚男和罗俊荷利用这些资料，以及其他相关资料，讲述了校园里15栋老建筑的故事。这本书展示了一所大学能给社会带来的影响。虽然荣杜易在成都停留的时间很短，但他的影响将通过这些建筑得以延续。上大学的地方对我们来说非常重要。那些有幸在美丽的建筑中接受教育的人，会在余生中留下关于校园的美好记忆。现在属于四川大学华西校区的老建筑，将深深地印在校友的脑海中，他们中的许多人虽然在远离成都的地方生活和工作着，但是会常常感受到母校与他们的关联。

赵安祝

赵丽勤

Richenda M. George

2022年4月5日

---

① 荣杜易到成都之旅的细节大部分来自于赵丽勤所拥有的荣杜易寄给家人和朋友的信。

② 改变了的地名，我们使用荣杜易可能使用的名称，在其第一次出现时括号里注明现在的名称。

③ 关于荣杜易的传记，请参见：Robson P (2014) *Fred Rowntree Architech: Some Notes on His Life and Buildings*. York: Newby Books.

PREFACE

On Saturday 1ˢᵗ February 1913 a 53 year old architect and his brother boarded a train from London to the port of Harwich at the start of what was an incredible adventure. His name was Fred Rowntree, and he had won the competition to design the buildings for a new university in Chengdu, China. This was the West China Union University, which was funded by four Christian mission organisations.

A direct flight from London to Chengdu takes under 11 hours. In 1913 the journey was rather more arduous.① From Harwich Fred, and his brother George, took a ship to the Hook of Holland and then a train to Peking (Beijing)② via Berlin, Warsaw, Moscow and Siberia. They arrived in Peking on February 14ᵗʰ and spent time in the capital meeting various people and visiting the sites that tourists still flock to today. On the 25ᵗʰ February they set off for Hankow (now part of Wuhan) by train, arriving on the 27ᵗʰ. They then took a boat, setting off on the 1ˢᵗ March sailing up the Yangtze by steamer to Ichang (Yichang) and then taking a kuaize (a small house boat) hauled by teams of trackers through the Yangtze gorges, at one point getting caught in a dangerous whirlpool, arriving on the 22ⁿᵈ March at Wan-Hsien (Wanxian). They set off across country, at some stages being carried in sedan chairs, staying in local inns or Buddhist temples. After spending time in Suining to observe local building styles and methods, they arrived in Chengdu on the 8ᵗʰ April.

It seems curious that after a journey that took more than 2 months, Fred and George spent less than a month in Chengdu before leaving on the 3ʳᵈ May. He spent that time exploring the site and its complications and, in a frenzy of meetings, getting to know the people on the ground, having discussions with the university, local government and church and missionary organisations. He also found time to draw up plans for the governor of the province for buildings in the Shaocheng Park and to discuss with him the concept of garden cities (a movement that Fred was closely involved with in Britain). He also helped with the design of other buildings, for example the Young Men's Christian Association (YMCA) building off Chunxi Road.

The journey back took even longer. He travelled with his brother, by wupan (small houseboat) to Ichang where they transferred to a steamer that arrived in Shanghai on the 27ᵗʰ May. His brother returned home by ship, but Fred travelled, via Japan, Hong Kong and Honolulu, to San Francisco (9ᵗʰ July) and then went by train to Toronto and New York (21ˢᵗ July) for talks with the missionary groups funding

1

the project. He finally returned home at the end of July 1913.

Fred never returned to China. But he continued to drive the project, receiving and sending from his offices at 11 Hammersmith Terrace in London countless letters, photographs and plans. Nothing was too small for his attention, he specified precise details of the decoration on the buildings. He also faced challenges, such as the discovery of old tombs that necessitated modification of the plans and negotiations with the local community. It is incredible to think that such a large and complicated project could be directed remotely, by letter, but Fred carried on working on the buildings of West China Union University until his death in 1927, when his son, Douglas Woodville Rowntree, took over.

While Fred was the architectural driver of the project it is of course important to remember others involved in the building. He worked with colleagues both in London (most notably his son Douglas and also Frank Osler who did many of the drawings) but more importantly the local people in Chengdu who had to turn his plans into reality. This included local managers and hundreds of craftsmen, workers and labourers who played their part in the enterprise. In addition other architects have contributed to the campus, 4 of the buildings described in this book were not designed by Fred.

Who was Fred Rowntree?[③] In 1913 he was a well-established architect, recognised for his work in the Arts and Crafts style that was popular in Britain. The Arts and Crafts movement built on the ideas and works of William Morris and others and emphasised the importance of hand-made, crafted, articles, in contrast to objects that were mass produced in factories. In architecture it encouraged the use of traditional building techniques, a simplicity of design, the use of local materials and a resistance to adhering to any one imposed architectural style. There was a political aspect to the Arts and Crafts movement, in addition to the concern about the impact of industrialisation, its followers were motivated by a desire to humanise both industry and the local setting people lived in and many had socialist tendencies.

Much of this Arts and Crafts heritage can be seen in his designs for West China Union University. His plans had been criticised by the Senate of the University who wanted 'modern western' buildings. But his adherence to the Arts and Crafts principles meant that he tried hard to understand local traditions, materials and building styles. He spent time on his travels investigating these, and from items he bought back from his travels we know he studied Chinese architecture. The result is seen in the buildings that are described in this book, which are a fusion of the British Arts and Crafts and traditional Chinese designs.

Fred Rowntree was born in the North Yorkshire seaside town of Scarborough to a family of grocers. His wider family included the Rowntree relatives who were based in York who are well known in Britain as chocolate makers, as well as philanthropists and social reformers. A very important aspect of Fred's life was that the family were Quakers. The Religious Society of Friends (the official name for Quakers) is a religious denomination that are characterised by not holding any doctrinal belief. While Quakers come from a Christian tradition they do not have a creed or set of beliefs that are required by worshippers. They have a strong commitment to pacifism, social responsibility and simplicity of life. In Quaker services (which do not take place in churches but in meeting houses) there is no priest or minister, nor any set order of service, but the Friends (as they call each other) sit in silence until someone is moved to say or share something with the others present.

His Quaker roots deeply influenced the work that Fred did, much of which was motivated by his sense of social responsibility. He was a prime mover in setting up Jordans, a model village that was built and run by its residents. He designed prefabricated homes for refugees in World War I, and helped set up a centre for rehabilitation of injured soldiers. One of his greatest interests was Hampshire House, a club that offered recreation, education and workshops for men and women from one of the most deprived areas of London. Many of the letters he sent back from China were to people involved in Hampshire House. In China he was similarly motivated by a desire to give people the opportunity of a university education that was relevant to their needs.

Those needs have changed over time, and so has the University. West China Union University has evolved, and the site is now a prestigious medical school, part of Sichuan University. This book shows how the buildings of the original university have been used in different ways, and some of the inspirational stories of people who have learned and worked in these buildings. The campus has continued to be relevant to many generations of students and staff, and it is gratifying to think of the impact they will have had locally, nationally and internationally.

Fred Rowntree was our great-grandfather (RMG) and great-great grandfather (AJTG). Andrew has been lucky to visit Sichuan University twice, and, as an academic, enjoyed seeing the research, learning and, in the associated hospitals, medical treatment that is taking place on the campus. Richenda briefly visited the campus in 1979, though access at that time was limited. We are sure that Fred Rowntree would be quietly pleased to see that the buildings that he designed are still functional, and appreciated for their beauty and architectural merit.

On his last visit Andrew was able, with his wife Professor Catherine Urch, to give to the University a collection of plans, photos, drawings and documents relating to the building of the campus that have been looked after for many years by Richenda. In this book, Qi Yanan and Luo Junhe have drawn on these sources, as well as others, to tell the story of 15 of the historic buildings on the campus. This book shows just how much difference a university can make to people and to communities. While Fred spent only a short time in Chengdu, we feel that his influence lives on through these buildings. The places that we go to school or university are very important to us. Those of us who have been lucky, and privileged, to be educated in beautiful buildings can carry the memories of them with us for the rest of our lives. The buildings that made up West China Union University, and which are now part of Sichuan University, will be imprinted in the minds of alumni, many of whom will be living and working far from Chengdu but will feel that connection with the place where they were educated.

Andrew JT George

Richenda M George

2022-4-5

① Details of Fred Rowntree's journey to Chengdu are largely drawn from letters he sent to family and friends that are in the possession of RMG.
② When place names have changed we have used the name that Fred Rowntree would have used, with the modern name in parentheses the first time.
③ For a biography of Fred Rowntree see: Robson, P (2014) *Fred Rowntree Architect: Some Notes on his Life and Buildings*. York: Newby Books.

前言 INTRODUCTION

　　成都南门外锦江河边的四川大学华西校区人称华西坝，有一百多年的历史。华西坝之前是古代名苑旧地，然而，到了清末，这里却成了夹杂着大量坟岗的农田。清末宣统二年（1910年），美国、加拿大和英国三国的5个基督教差会在此置地创办了华西协合大学，随后城南这片地区就有了新的称呼——华西坝，现在成都市的公交和地铁在此地的站名就叫华西坝。

　　The West China Campus of Sichuan University by the Jinjiang River, outside the south gate of Chengdu, has a history of more than one hundred years. The Campus ground was previously an ancient famous garden. However, in the late Qing Dynasty, it was a farmland dotted with a lot of graves. In 1910, five Christian missions from the USA, UK and Canada bought land here, and established West China Union University. Then the area is called Huaxiba by local citizens. Huaxiba is so well known that now, the bus stop and subway station here were named Huaxiba Station.

　　19世纪末，已有美国、英国和加拿大三国的8个基督教差会来四川传教，他们中的一些传教士认识到教育、医疗对于传教活动具有促进作用，因而这些差会在四川各地开办了众多的幼稚园、小学和中学等教育机构以及医院。特别是1905年清政府废除了科举制度，大力开办新学、培养新式人才，这让传教士们的办学有了更宽松的发展空间。同年，基督教教会成立临时管理部负责大学的筹建工作。

　　In the late 19th century, eight Christian missions from the USA, UK and Canada came to Sichuan to preach. Some of the missionaries realized that education and medical care could promote missionary activities. So these

missions established many hospitals and educational institutions in Sichuan, including kindergartens, primary schools, middle schools, and hospitals. Especially in 1905, the Qing Government abolished the imperial examination system, vigorously established new schools and trained new talents, which also gave the missionaries a more relaxed development environment to run schools. In the same year, the church established a temporary management department to prepare the University.

外国教会来华开办学校和医院的初衷是为了传教，但就像"蜜蜂本意是觅食，但它传播了花粉"一样，传教士在传教的同时，也将西方先进的科学技术传播到成都，中国文化也通过他们传播到西方社会。因而华西坝这一中西文化交汇之地，不仅是西部近代高等教育的发祥地，而且也是成都具有特殊文化的区域之一。成都文化名人流沙河说，华西坝是成都的五大文化标志地之一（老少城、华西坝、草堂、武侯祠和春熙路）；宽窄巷子和华西坝是成都一中一洋两种最典型的文化的代表。

The purpose of those foreign missions, which established schools and hospitals in China, is preaching. However, "bees intend to collect pollen, but they also spread the pollen." While preaching, missionaries also spread advanced western science and technology to Chengdu, and Chinese culture also spread to the western society by the missionaries. So Huaxiba, confluence of Chinese and western cultures, is the birthplace of modern higher education in western China, also is one of Chengdu's areas with special culture . As famous local cultural master Liu Shahe said, Huaxiba is one of the five major cultural landmarks in Chengdu(The Lesser City of Chengdu, Huaxiba, Du Fu Thatched Cottage, Wuhou Shrine and Chunxi Road). Kuan-Zhai Alley represents Chinese style and Huaxiba represents western style. Both are the two most typical cultural symbols of Chengdu.

华西协合大学创办初始，校园应该怎样规划？建筑物是中式还是西式？首任校长美国人毕启曾回忆道："我们采用了与中国传统建筑和谐一致的大学建筑风格，我们开创了一种现在已经在中国其他基督教大学中普遍采用的建筑类型。"

At the beginning of the West China Union University was founded, how should the campus be planned? Is the building style Chinese or Western? The first university president, American Dr. Joseph Beech recalled, "In adopting a style of collegiate architecture in harmony with Chinese traditional architecture, we pioneered in a type which has become general in other Christian universities in China."

1912年，校方在加拿大、美国和英国等国专门举办了校园建筑设计竞标赛，英国建筑师荣杜易综合了中西方建筑工艺元素的方案被选中。他对华西协合大学的总体规划和设计在当时的中国学校中是独一无二的，所以后来其他一些中国大学也采用了这种设计风格。这些建筑都具有大屋顶、青砖黑瓦、斗拱飞檐等中国传统建筑的风格，且建筑布局呈中轴对称，而屋顶上的塔楼、烟囱、雉堞，却明显显示出西方建筑的风格。

In 1912, the University authorities held special campus architectural design competitions in the USA, UK and Canada, and the scheme of British architect Fred Rowntree was selected. The scheme integrated the elements of Chinese and Western architectural craft, and the overall plan and design of West China Union University was unique among Chinese schools at that time. Some other Chinese universities later adopted the form of this design. These buildings all have the style of traditional Chinese architectures, such as large roof, black bricks and tiles, bucket arch eaves, and the architectures symmetrical layout the axis line, while the towers, chimneys, crenelations on the roof clearly show the western architectural style.

校长毕启写了一篇文章，对荣杜易设计的大学校园是这样评价的："他开创设计了一种既有最华丽中国建筑的元素，又有西方建筑的稳定性的一种建筑风格。他创作的建筑方案不仅获得了校园建筑设计竞赛大奖，而且在他访问成都时，一旦他向当地名流政要们展示他的绘图作品，这些中国人会立刻被他的作品所吸引。当我们把提交的三份参与竞标的大学设计方案向一群中国杰出的绅士展示时，他们在并不知道是哪一套方案赢得了这个奖项的情况下，仍然选择了荣杜易设计的这一套结合了中国建筑之美的理念的方案。这也证实了董事会选择荣杜易成为学校建筑设计师是正确的。刚开始我们校园里的建筑没有统一的风格，杂乱无章，后来荣杜易参与校园建筑设计后，为我们提供了独特的具有东方风格的学校建筑方案，使得我们这所大学的校园在中国的学院中独一无二。"

Dr. Beech wrote an article talking about the university campus designed by Mr. Rowntree, "He developed a type of architecture that combined the noblest elements in Chinese architecture with the stability of the West, producing a style of building which not only won the award, but at once captivated leading Chinese to whom his drawings were shown upon his visit to Chengdu. Three sets of plans submitted in competition were spread before a group of eminent Chinese gentlemen. They were not told which set of plan had won the award. They chose that set which incorporated their own ideas of beauty in architecture and thus confirmed the action of the Board of Governors in making Mr. Rowntree the architect. It was first our dire poverty and later Mr. Rowntree's coming that saved us from 'hodgepodge' and gave us the superb Oriental collegiate architecture that has made the campus of this University unique among China's Colleges."

百年来这些老建筑不仅为培养学生提供了教学场所，而且也见证了学校的发展历程，它们是我们这所学校和城市的文化基因。

For the past hundred years, these architectures have not only provided teaching places for students, but also witnessed the development of the university. They are the cultural genes of our university and the city.

当年修建这些教学楼的资金主要是由美籍校长毕启募集而来的，所以当初每一栋楼都以捐赠者的名字命名。比如怀德堂、万德堂等。到了20世纪50年代初，学校用数字给华西坝上的教学楼重新命名，就是现在的第一教学楼、第二教学楼……总共命名了8座教学楼，而大量的居家小洋楼却陆陆续续被拆除了，现在仅存3栋小洋楼。

The funds for the construction of these buildings were mainly collected by Dr. Beech (American), the president of the university. As a result, each building was originally named after its donor, such as the Whiting Memorial

前言

Administration Building, the Vandeman Memorial Hall, and so on. By the early 1950s, the teaching buildings in Huaxiba were renamed with numbers, such as the First Teaching Building, the Second Teaching Building...Totally eight teaching buildings were numbered. A large number of small houses had been demolished one after another, and now only three remained.

2013年5月,学校对校区的建筑楼宇重新命名,其名称更具备中国文化特色,比如启德堂、志德堂等。同年5月,华西坝上的12栋老建筑中的8栋被国务院核定为"全国重点文物保护单位",这就意味着华西坝的老建筑成了"国宝"。

In May, 2013, again the University authorities renamed the buildings on the campus. The new names presented more Chinese historical imprints. In May of the same year, 8 of the 12 old buildings in Huaxiba were approved as "National Key Cultural Relic Protection Units" by the State Council. It means that the 8 old buildings of Huaxiba are "national treasure".

本书通过展示英国建筑师荣杜易设计的华西坝建筑的部分手稿、设计图纸,以及最终落成的现存楼宇今昔照片的对比,同时讲述这些建筑里蕴藏的人文历史故事,让它们更具有生命力和文化内涵,让大家感受华西坝百年老建筑及其历史文化。

This book shows some manuscripts and design drawings of Huaxiba buildings designed by British architect Rowntree, as well as photos of the past and present of existing old buildings. At the same time, it tells the humanistic and historical stories of these buildings, so as to endow these buildings with more vitality and cultural connotation, and let everyone know the century old buildings and their historic culture of Huaxiba.

笔者关注华西坝老建筑已有20多年,其间收集了许多相关的老照片与资料,并且一直坚持拍摄华西坝现存的老建筑。我们还从耶鲁大学网站上收集到不少珍贵的华西坝的老照片。2007年,耶鲁大学善本藏书馆馆员玛莎·斯莫利来信允许我们下载并使用该网站上有关华西坝的老照片。

The authors have been paying attention to the old buildings of Huaxiba

Yale University

Martha L. Smalley
Special Collections Librarian/
Curator of the Day Missions Collection
Divinity School Library
409 Prospect Street
New Haven, Connecticut 06511-2108

Campus address:
212D SDQ
Telephone: 203 432-5289
Fax: 203 432-3906
Email: martha.smalley@yale.edu

November 1, 2007

Qi YaNan
Chengdu city
Sichuan Province
PRC CHINA

Dear Qi YaNan,

Thank you for sending us a copy of your book related to West China Union University. We are pleased to add it to our collection.

You have our permission to download photographs related to WCUU from the Yale Divinity School web site for your work.

Sincerely,

Martha L. Smalley
Special Collections Librarian

这就是2007年11月1日，耶鲁大学善本藏书馆馆员玛莎·斯莫利寄来的允许作者下载并使用该图书馆网站上有关华西坝的老照片的信。

This is the letter came from Martha L. Smalley, a Special Collections Librarian of Yale University, on November 1st, 2007, allowing the author to download and use old photos related to West China Union University on the library's web site.

for more than 20 years, during which time we have collected many old photos and materials, and have been photographing the existing old buildings in Huaxiba. We also collected many precious old photos of Huaxiba from the Yale University website. In 2007, a letter from Martha L. Smalley, the librarian of Yale Library allowed us to download and use the old photos of Huaxiba, on the site.

本书的成书还得益于赵安祝提供的其高祖父荣杜易当年设计华西协合大学的建筑图纸等历史文献资料。这本书也是笔者戚亚男于2021年在成都市哲学社会科学重点研究基地成都史学与成都文献研究中心获得的研究"华西坝老建筑里的历史人物"课题项目（编号为CLWX21006）的成果。

This book is completed thanks to the historical documents provided by Dr. Andrew George, such as architectural drawings of the West China Union University designed by his great-great-grandfather Rowntree. In 2021, the author Qi Yanan won the research project of *The Historical Figures in the Old Buildings of Huaxiba* (No. CLWX21006) at Chengdu History and Documentation Research Centre of Chengdu Key Research Base of Philosophy and Social Sciences. This book is also the result of this research project.

戚亚男

2022年4月15日于华西坝

Qi Yanan

Huaxiba, April 15, 2022

目 录

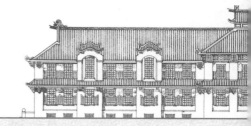

壹 / 华西坝的总建筑设计师——荣杜易
Fred Rowntree, the Chief Architect of Huaxiba 1

贰 / 华西坝建楼宇善款的募集者——毕启校长
President Joseph Beech, the Fund Raiser of Huaxiba Buildings 17

叁 / 走进华西坝——校门
Walk into the University Gate of Huaxiba 23

肆 / 华西坝的标志——钟楼
The Coles Memorial Clock Tower, the Landmark of Huaxiba 39

伍 / 华西校区行政楼——怀德堂
The Whiting Memorial Administration Building of Huaxiba Campus 57

陆 / 西部图书馆与博物馆肇始——懋德堂
The Lamont Library and the Museum Building, the Origin of West China Library and Museum 71

柒 / 自然科学的圣地——嘉德堂
The Atherton Building for Biology and Preventive Medicine, the Holy Land of Natural Science 85

捌 / 抗战时期的建筑文物——懿德堂
Stubbs' Memorial Chemistry Building, the Building of the Counter-Japanese War Age 97

玖 / 因市政建设而移建的学舍——万德堂
The Vandeman Memorial Hall, Once Relocated Due to Municipal Construction 109

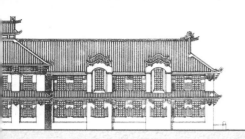

CONTENTS

拾 / 汇聚众多中国文化学者的文学院——广益大学舍
The Friends' College Building, the Liberal Art School that Gathers Many Chinese Scholars......121

拾壹 / 中西文化的交汇处——合德堂
The Hart College, Where East Culture Meets West......135

拾贰 / 教育学院——育德堂
The Cadbury School of Education Building......149

拾叁 / 西部医学教育的发祥地——启德堂
The Medical and Dental College Building, the Cradle of West China Medical Education......169

拾肆 / 抗战催生的大学医院——华西协合大学医院
West China Union University Hospital, Which Was Built During the Counter-Japanese War......195

拾伍 / 开男女同校先河——女子大学舍
The Women's College House, the First Coeducation......213

拾陆 / 华西坝上的教职员住宅——小洋楼
The Faculty and Staff Residence, Small Foreign Style Houses in Huaxiba......229

拾柒 / 中加两国人民友谊的见证——志德堂
The Canadian School in Huaxiba, Witnessing the Friendship Between Chinese and Canadian People......253

华西坝老建筑的前世今生
The Stories of the Historic Architecture of Huaxiba

华西坝的总建筑设计师

荣杜易

Fred Rowntree, the Chief Architect of Huaxiba

壹·华西坝的总建筑设计师——荣杜易

　　1912年，华西协合大学创办者在加拿大、美国和英国等国专门举办了大学校园建筑的设计竞标赛，英国荣杜易父子的建筑公司一举夺魁并拿到华西坝校园建筑的设计权。次年，53岁的荣杜易专门从英国来成都实地考察所要设计的建筑的环境和当地的文化特色。他从俄国进入中国，先到北京，特意参观了故宫等具有显著中国传统风格的建筑，后于4月初来到成都。在入川的路上，荣杜易看到了西南地区建筑的特有形式，如穿斗式屋架、屋顶的脊饰等，这些具有川西风格的建筑形式，在他日后为华西协合大学设计的建筑物中都有具体的体现。

　　In 1912, the founders of West China Union University held special campus architectural design competitions in Canada, the USA, and the UK. The British Fred Rowntree and sons architectural firm won the competition, and the design right of Huaxiba campus architecture. The following year, Mr. Rowntree, in his 50s, came to Chengdu from England to investigate the environment and local cultural characteristics of the buildings to be designed. He entered China from Russia, went to Beijing to visit the Palace Museum and outstanding buildings with Chinese style first, and then went to Chengdu in early April. On the way into Sichuan, He saw the unique architectural techniques in southwest China, such as column and tie frame and roof ridge decoration. These forms of western Sichuan buildings were embodied in the buildings designed for West China Union University in the future by Mr. Rowntree.

左图：英国建筑师荣杜易。赵安祝提供。
British architect Rowntree. Andrew George provided.

右图：1913年荣杜易来成都实地考察，为华西协合大学校园建设规划时，就是通过当时对他而言是唯一的方式——乘船逆长江入川。赵安祝提供。
1913, Mr. Rowntree came to Chengdu to investigate and design West China Union University construction. It was the only way for him to take boat upstream the Yangtze River into Sichuan. Andrew George provided.

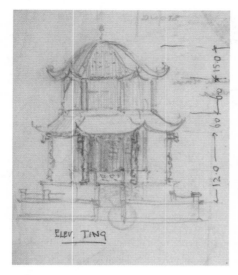 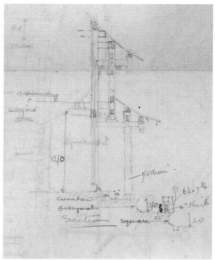

最初规划设计的华西协合大学占地近900多亩（1亩≈667平方米），校园的朝向没有按中国传统习惯建成坐北朝南的格局，而是采取坐南朝北格局，这是因为这块校址的南边是城郊，而北边临府南河，就在河的对岸是这座城市。如果坐北朝南，背靠一条河，不仅不符合中国传统建筑的风水理论，而且也不利于大学和城市的交往，以及今后学校的发展。基于这些因素，建筑师荣杜易把大学的校门朝向南河、朝向城区，也就是向北面开门。

In the original plan, West China Union University covered an area of nearly 0.6 km². The campus was not the pattern of facing towards the south according to the traditional Chinese habits, but the pattern of facing towards the north, because the south of the campus was the suburbs and the north was Funanhe River. Across the river is the city. It was against the Fengshui theory of traditional Chinese architecture to set the university gate not face toward the river, nor was beneficial for communication and development of the university. Based on these factors, the architect Rowntree designed the university gate faces the South River and the city in north.

荣杜易在成都仅考察了不到1个月就回国了，而他设计的华西坝的建筑却有众多的中国传统建筑元素，他又是如何掌握这些中国传统建筑元素并且应用到华西坝建筑的设计中的呢？原来荣杜易回国后，通过书信往来对华西坝建筑的设计以及修建进行指导。就在荣杜易离开成都两个多月且还没有回到英国

左图：莫里森通过邮政提供给荣杜易他手绘的一座八角楼亭的立面图手稿。赵安祝提供。
By a letter, Mr. Morrison provided Mr. Rowntree the manuscript of the elevation of an octagonal pavilion painted by him. Andrew George provided.

右图：莫里森提供给荣杜易他手绘的一座八角楼亭的剖面图手稿。赵安祝提供。
By a letter, Mr. Morrison provided Mr. Rowntree the cross section manuscript of an octagonal pavilion painted by him. Andrew George provided.

时，他在四川遂宁的朋友莫里森先生，于1913年7月1日就给他写信，告诉荣杜易他专门爬上正在维修的寺庙屋顶，观察当地建筑民工如何修建脊兽："我很幸运，能够如此近距离看到屋脊上的那些脊兽小动物，特别是鱼，这些脊兽都是用黏土做的。脊兽鱼的重量巨大，是在屋脊的末端做成整条的鱼形状。""山那边庙里的小门上有一幅鱼的小画像，你会很喜欢的。"莫里森随信还提供了他手绘的一座八角楼亭的立面图、剖面图和平面图手稿。通过这种方法，荣杜易更进一步了解到中国传统建筑艺术，并融入到他设计的华西坝建筑方案中。

The architect Rowntree spent less than a month in Chengdu before returning home. But the architecture of Huaxiba had many Chinese architectural elements. How did he master these Chinese architectural arts and applied them to the design of Huaxiba? It turned out that after Mr. Rowntree returned to the UK, he maintained the guidance of the design and construction of the Huaxiba buildings through letters. More than two months after leaving Chengdu, before he returned to Britain, on July 1, 1913, his friend Mr. J. H. Morrison in Suining, Sichuan Province, wrote to Mr. Rowntree saying that he purposely climbed up the roof of the temple that was being repaired to see how the local construction workers built the ridge beasts. "I was fortunate also in being able to see those little animals and fishes close to. All are made of the clay. The fishes are of enormous weight and are made in one piece with the end of ridge." "There is a small sketch of the fish on the little gateway inside the temple on the other side of the hill which you would like." Mr. Morrison also provided the elevation, section, and floor plan manuscripts of his hand-painted octagonal pavilion in the letter. Through this method, Mr. Rowntree learned more about Chinese architectural art and integrated it into his design of Huaxiba architecture.

1924年，荣杜易在英国《建筑师》杂志上这样写道："在开始建设之前，资深合作伙伴参观了现场，他们不仅与理事会成员，还与该省的三位官员讨论了各种问题。这些中国官员赞同建筑设计应该具有中国特色的建议，并且礼貌地补充说，如果按照这种风格修建学校，今后他们可以复制！我们努力保持过去历史遗留下来的形式、纹理和颜色，明智、恰当地使用该国能够提供的材料和建筑方式，并使它们适应现代需求。"

In 1924, Mr. Rowntree wrote in the British magazine *The Builder*, "Before commencing building operations, the senior partner visited the site and discussed the various problems, not only with the members of the Senate, but also with the three leading Chinese statesmen of the Province, who welcomed the suggestion that the design of the buildings should be Chinese in character, politely adding that if they were carried out in that spirit they could copy them! An endeavour has been made to maintain the forms, texture and colouring handed down from past history, and to adapt these to modern requirements, with the judicious and harmonious use of such materials and forms of construction as the country can best supply."

荣杜易对成都华西协合大学的总体规划和设计在当时的中国学校中是独一无二的，所以后来其他的一些中国大学也采用了这种设计方案。2013年5月，华西

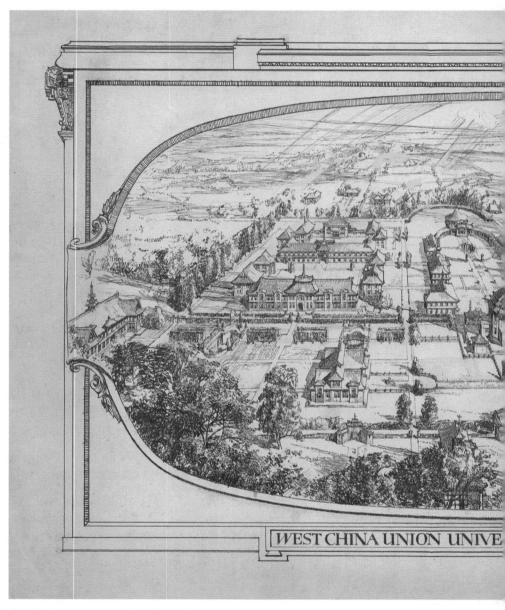

荣杜易设计的华西协合大学校园鸟瞰图手稿。赵安祝提供。
A bird's-eye view of the campus of West China Union University manuscript designed by Mr. Rowntree. Andrew George provided.

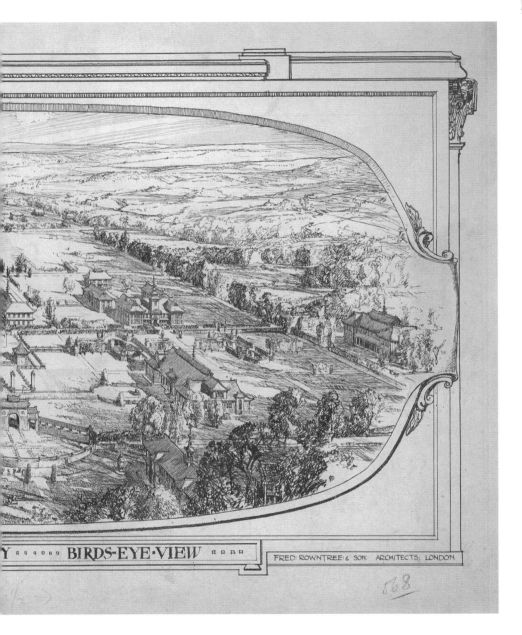

BIRDS·EYE·VIEW   FRED·ROWNTREE·&·SON·ARCHITECTS·LONDON

壹 · 华西坝的总建筑设计师——荣杜易

坝上被国务院核定为"全国重点文物保护单位"的8栋老建筑,其中有7栋是荣杜易设计的。

Mr. Rowntree's overall plan and design of West China Union University in Chengdu were unique among Chinese schools at that time, so some other Chinese universities later adopted the style of this design. In May, 2013, eight old buildings in Huaxiba were approved as "National Key

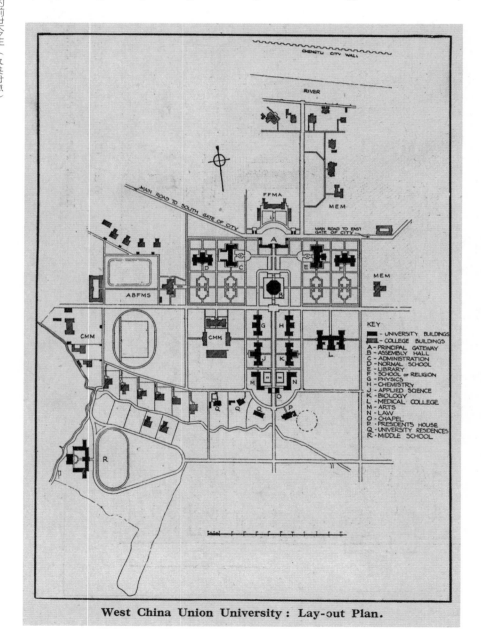

West China Union University: Lay-out Plan.

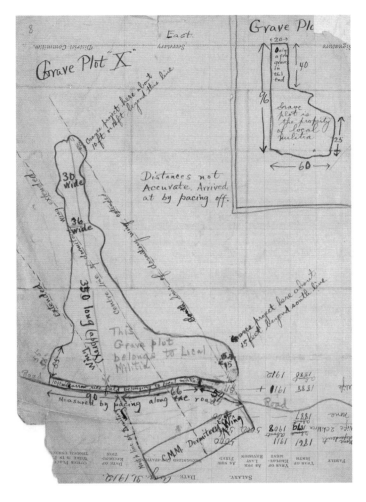

左图：荣杜易设计的校园建筑规划图。资料来源：1924年英国《建筑师》杂志。
Campus building planning map designed by Mr. Rowntree. Source: 1924 British magazine *The Builder*.

右图：荣杜易在一张废弃的1912年工资登记表上手绘的校园平面图。赵安祝提供。
A hand-painted campus floor plan on an abandoned Salary Registration Form of 1912 by Mr. Rowntree. Andrew George provided.

Cultural Relic Protection Units" by the State Council. Among them seven buildings were designed by Rowntree.

令人遗憾的是，荣杜易回英国后，直到1927年去世前都没有机会回到华西坝再看看他设计的大学校园。

It is a pity that after Mr. Rowntree returned to Britain till his death in 1927, there is no chance for him to return to Huaxiba to enjoy his masterpiece.

当年英国权威杂志《建筑师》为荣杜易的去世发布的讣告上说："1912年，他被任命为位于四川省成都市的华西协合大学的建筑师，该校建筑面积很大，至今仍在建设中。"

2018年3月，赵安祝捐赠他的高祖父荣杜易当年设计的华西坝的建筑图纸等历史文献资料给四川大学。王允保摄于2018年。

In March 2018, Dr. George, a descendant of Mr. Rowntree, the designer of the old Huaxiba buildings, donated the architectural drawings for Huaxiba and other historical documents of his great-great-grandfather's to the University. Photo by Wang Yunbao in 2018.

The Builder, a leading British magazine released an obituary of Mr. Rowntree, "In 1912 he was appointed architect to the West China Union University at Chengtu Szech-uan, the buildings of which cover a large area and are still in progress."

"他是一个实践自己理想的人，也是一个优秀的商人。他为自己立下的目标而努力工作。他有相当大的能力和优秀的品格，他强壮，温和，谦虚，善良，实际，头脑清醒。"

"He was a man who lived up to his ideals, and also an excellent man of business. He worked hard for those things he set his hand to. He had considerable abilities, and sterling character; he was strong, but gentle, modest, kind, practical, and clear-headed."

2018年3月，满园春色的华西坝迎来了一位特殊的英国友人——英国布鲁内尔大学副校长赵安祝，他的高祖父就是荣杜易。赵安祝此次来四川大学的主要目的就是捐赠他的高祖父荣杜易当年设计华西坝的建筑图纸等历史文献资料。

In March, 2018, Huaxiba welcomed a special British friend, Vice President of Brunel University Dr. Andrew George, whose great-great-grandfather was Fred Rowntree. The main purpose of his visiting to the University was to donate the architectural drawings and other historical documents of Huaxiba designed by his great-great-grandfather Fred Rowntree in that year.

捐赠后的第二天下午5点，赵安祝来到华西坝寻找高祖父当年设计的老建筑，笔者受邀去给赵安祝讲解这些老建筑。我们首先参观了华西坝的标志性建筑——钟楼，赵安祝拿出随身携

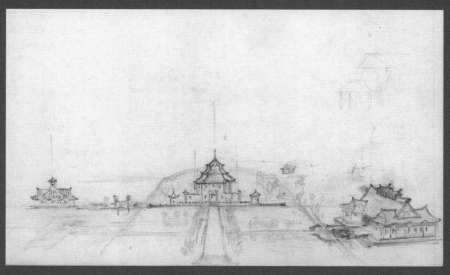

荣杜易手绘的聚会所草稿。赵安祝提供。
A hand-painted Assembly Hall draft by Mr. Rowntree. Andrew George provided.

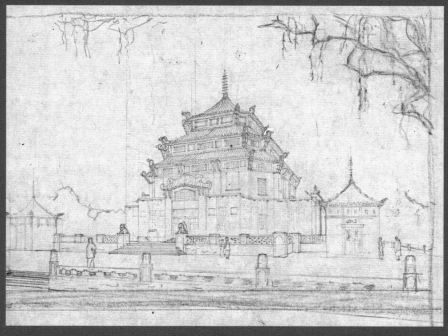

荣杜易设计的一款聚会所手稿。赵安祝提供。
An Assembly Hall manuscript designed by Mr. Rowntree. Andrew George provided.

左图：1926年竣工的钟楼，屋顶是中国传统建筑形式歇山顶，在歇山顶上又建造了一个尺寸较小的四角攒尖顶的方形塔。赵安祝提供。
The Coles Memorial Clock Tower, completed in 1926, with a roof in the traditional Chinese architectural form, and a smaller square four-corner pyramidal spire was built on the top. Andrew George provided.

右图：20世纪50年代钟楼维修过程中，中国建筑设计师古平南对钟楼塔顶做了很大的改动，把原来的歇山顶改为中国传统建筑屋顶形式中最为复杂的十字脊歇山顶，这样从东、南、西、北四面看过去钟楼屋顶都是一样的。咸亚男摄于2021年。
During the repair of the clock tower in the 1950s, Chinese architect Gu Pingnan made great changes to the top of the tower, changing the original rest top to the most complex cross ridge rest top in the traditional Chinese architectural roof form, so that look from east, south, west and north, the tower roof is the same. Photo by Qi Yanan in 2021.

带的平板电脑，从里面调出两张钟楼老照片，一张是正在建筑中的，另一张是已经落成的。老照片里钟楼的屋顶是中国传统建筑形式歇山顶，在歇山顶上又建造了一个尺寸较小的四角攒尖顶的方形塔，而此时矗立在赵安祝面前的钟楼却与老照片里的钟楼不同。笔者告诉赵安祝，现在的钟楼外形与过去的钟楼不同，那是因为在20世纪50年代钟楼经过维修，中国建筑设计师古平南对塔顶做了很大的改动，把原来的歇山顶改为中国传统建筑屋顶形式中最为复杂的十字脊歇山顶，较小的四角攒尖顶的方形塔也做了相应的修改，使其看起来与钟楼下半部分更一致，更具有中国建筑的特色。笔者的解释，让赵安祝相信了此钟楼就是彼钟楼。

At 5 p.m. of the next day, Dr. George came to Huaxiba to find the old buildings designed by his great-great-grandfather Rowntree. The author was invited to interpret these old buildings to him. First we

左图:走到钟楼前的小桥上,看着对面的钟楼,赵安祝立刻邀请笔者与他以钟楼为背景合影留念。左起赵安祝、咸亚男、罗俊荷。咸亚男提供。

Walking to the small bridge in front of the clock tower, Dr. George immediately invited we to take a photo with him with the clock tower as the background. From left Dr. George, Qi Yanan, Luo Junhe. Provide by Qi Yanan.

右图:赵安祝手持1915年拍摄的怀德堂奠基仪式的老照片在奠基石前合影。咸亚男摄于2018年。

Dr. George took a photo in front of the cornerstone. In his hand is an old photo of the Whiting Memorial Administration Building foundation stone laying ceremony in 1915. Photo by Qi Yanan 2018.

visited the landmark clock tower of Huaxiba. Dr. George called up two old photos of the clock tower from his iPad, one under construction and the other completed. The roof of the clock tower in the old photo is traditional Chinese architectural form of the rest top, and there is a smaller quadrangle spire on the rest top. Apparently, the clock tower in front of us is different from the clock tower in the two old photos. Why? The author told him that was because of the tower maintenance in the 1950s, Chinese architect Gu Pingnan made great changes to the top of clock tower, changing the original rest top to the most complex cross ridge rest top in the traditional Chinese architectural roof form, modifying the smaller quadrangle spire accordingly, made the clock tower looked the upper half more consistent with the lower half and more characteristic of Chinese architecture. The author's explanation convinced him that this clock tower was the one that Rowntree designed.

钟楼东边的华西药学院的教学楼引起了赵安祝的注意,笔者告诉赵安祝,这栋教学楼就是万德堂,1960年因成都市修建人民南路,将该楼迁建于钟楼旁,基本还原其本来面目,但与原建筑不同的是屋顶上中式攒尖顶圆亭没有了。赵安祝随即调出平板电脑里的万德堂的老照片与其进行比对,的确,除了没有万德堂老照片屋顶上的一座两层圆亭,我们面前的教学楼与老照片上的建筑基本一致。随行的赵安祝夫人还特别问了此楼是否是用拆下的原来的建材重建的,得到肯定回答后,夫妇双方都大为赞赏。

The teaching building of West China School of Pharmacy, Sichuan University to the east of the clock tower attracted the attention of Dr. Andrew George. The author told him that the teaching building was the Vandeman Memorial Hall. In 1960, due to the construction of the South Renmin Road in Chengdu city, the building was moved next to the clock tower, basically to restore its original appearance, but different from the original building, the Chinese-style saving spire round pavilion on the roof was gone. Dr. George then called out the old photos of the Vandeman Memorial Hall in the iPad and compared them. Indeed, except for a two-story round pavilion on the roof of the old photo of the Vandeman Memorial Hall, the teaching building in front of us is no different from the buildings in the old photos. Accompanying Mrs. George also asked whether it was rebuilt with the removed original building materials. After receiving a positive answer, the couple were greatly appreciated.

　　随后我们去参观怀德堂（华西校区办公楼），这座建筑是华西坝上老建筑群中较早修建的，它是荣杜易的得意之作。当笔者带赵安祝去看怀德堂刻有修建时间的奠基石时，赵安祝很是兴奋。他从平板电脑里调出一张老照片进行比对，老照片是怀德堂1915年举行奠基仪式时拍摄的，照片显示在建筑工地东头一角的地基上用几块布搭了一个简易的棚子，里面有一面写着"华西协合大学"的旗帜，棚子前面摆放了几排桌椅，一块刻有"A.D. 1915""中华民国四年"的奠基石镶嵌在正在修建的大楼墙体里，与眼前所见的奠基石一模一样。这次赵安祝自己不拍照了，他把手里的平板电脑中的老照片放大到能看见奠基石上的"A.D. 1915"字样，自己站在奠基石前面，手持平板电脑，让我们给他拍照，与不同时期的同一块奠基石合影，多么神奇！这张合影对于他来说有十分重要的意义。穿越了百年时光，他把自己高祖父所留下的有这栋建筑奠基石的老照片带到了有"A.D.1915"奠基石的怀德堂前。奠基石的照片与实物跨越四个世代的时间距离，在同一个空间零距离接触，并同镜入照，此时赵安祝是怎样的一种心情，我们不得而知。两张穿越时代的照片弥足珍贵，将其间的故事深深地烙印在了历史的长河中。

怀德堂1915年举行奠基仪式时拍摄的照片，刻有"A.D. 1915"字样的奠基石矗立在大楼基座上，人们正准备将奠基石镶嵌在正在修建的大楼墙体里。赵安祝提供。

In 1915, the Whiting Memorial Administration Building foundation stone laying ceremony. The foundation stone inscribed "A.D. 1915" stood on the platform of the building and soon would be embedded in the walls of the building. Andrew George provided.

Then we went to visit the Whiting Memorial Administration Building. It was built earlier than many other old buildings in Huaxiba, and it was a proud work by Mr. Rowntree. When the author introduced Dr. George the

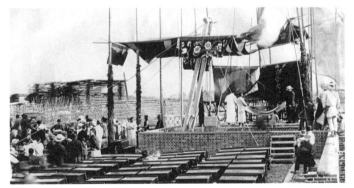

foundation stone of the Whiting Memorial Administration Building, he was very excited, for the foundation stone identical to the one on an old photo he called out from his iPad. The old photo was taken in 1915, when the building was constructing. The photo shows a simple shed built with several pieces of cloth on the foundation at the eastern corner of the construction site, with a flag reading "West China Uniton University" there, a few rows of tables and chairs placed in front of the shed, and a cornerstone engraved with "A.D. 1915" and "the 4th year of the ROC" embedded in the wall of the building under construction, identical to the cornerstone seen in front of the building. This time Dr. George would not take pictures. He magnified the old photos in his iPad so that the phrase "A.D. 1915" on the foundation stone was big enough to be seen clearly. Then he stood in front of the foundation stone and holding the iPad with the old photo, and let me take a photo of him. How amazing it is to take a photo with the same cornerstone of different periods! This photo means a lot to him. After a hundred years, he brought the old picture of the cornerstone of this building left by his great-great-grandfather to the front of the Whiting Memorial Administration Building with the cornerstone of "A.D.1915". There is a time distance of four generations. The cornerstones are in contact with each other in the same space at zero distance, and appear in the same photo. At this time, we do not know what kind of mood Dr. George has. But the two photos that go through the times are precious, and the stories in them are deeply imprinted in the long river of history.

就这样我们在校园里慢慢地走着，一边对照着老照片观看老建筑，一边交谈着。赵安祝说，1913年他高祖父来成都时，乘船逆江入川，在三峡遇险，船失去控制，在原地急速打漩往下沉，高祖父以为自己活不成了，没想到最后被船工救上了岸。接着他又说，高祖父非常珍惜他为华西坝设计的建筑原始手稿，一直都细心收藏着。荣杜易家人的住宅也有一些华西坝建筑风格。笔者发现赵安祝对他高祖父设计的建筑的细节特别感兴趣，这些建筑很多地方带有中国元素，如建筑屋顶上的各种脊兽，他都要与他平板电脑里的老照片进行比对，还随时用手机拍照。

We walked slowly on the campus, watching the old building when compared with the old photos and talking. Dr. George said when Mr. Rowntree came to Chengdu, Sichuan, by boat in 1913, he encountered great distress at the Three Gorges. The boat lost control and swirled rapidly and sank down. Rowntree thought he could not survire. Fortunately he was finally rescued by the boatman. Then he said Rowntree cherished the original architectural manuscript he designed for Huaxiba and had always carefully retained. The family residence of Mr. Rowntree also has some style of Huaxiba architecture. Dr. George was particularly interested in the details of the architecture which Rowntree designed, and although many places had Chinese elements, such as various ridge animals on the building roof. He compared to the old photos on his iPad, and take photos with his cellphone.

参观近两个小时后天色已晚，赵安祝意犹未尽，我们相约下次赵安祝再来华西坝时，我们再一起去参观华西坝上的他高祖父设计的其他老建筑、摆老建筑的龙门阵[①]。

Night falls. Dr. George wanted to see more old buildings, but time was limited. We would like to make an appointment to meet in Huaxiba, to visit the other old buildings designed by his great-great-grandfather.

---

①龙门阵：方言，前加一个摆字，意为聊天、讲故事。

华西坝老建筑的前世今生
The Stories of the Historic Architecture of Huaxiba

贰

华西坝建楼宇善款的募集者

毕启校长

President Joseph Beech, the Fund Raiser of Huaxiba Buildings

建校之初，为了让大学有合格的生源，1908年，大学的临时管理部把市区里教会办的三所中学迁到华西坝来，成立了华西高等预备学堂。学堂在购置的土地上修建了两三座简易的平房，这种四川最常见的平房为穿斗式木结构，墙体为竹编夹泥墙，外抹白灰，屋顶为灰色瓦。1910年，从华西高等预备学堂的100位学生里挑选了11位进入华西协合大学学习，这是华西协合大学的第一届学生。这些简易的平房就是当时大学的教学场所，此时大学的建设方案还未确定。直到两年后，英国建筑师荣杜易中标成为华西协合大学的总设计师，为校园设计建设方案。

At the beginning of university establishment, in order to let the university admit high quality students, in 1908, the Temporary Board of Management of the university moved the three middle schools run by the urban church to Huaxiba and established the Huaxi Higher Preparatory School. Two or three simple bungalows were built on the land purchased. Such most common bungalow in Sichuan was a column and tie type wooden structure, with bamboo wall mud covered by white ash(lime), and a grey tile roof. In 1910, 11 were selected from the 100 students in Huaxi Higher Preparatory School as the first students to attend West China Union University. These simple bungalows were the teaching places of the university at the time. At this time, the construction plan of the university has not yet been decided. Two years later, the British architect Fred Rowntree won the bid to become the chief architect of West China Union University, designed permanent buildings for the campus.

华西协合大学首任校长毕启。图片来源：四川大学档案馆。
Dr. Joseph Beech, the first president of West China Union University. Photo source: Sichuan University Archives.

建校之初，从校舍之修建、教学设备之购置到师资之聘任等都离不开巨大的经费开支。对于一所私立大学，这一切费用都要自己筹措，因而从建校之初校长毕启就为了维持学校的正常运转在国内外到处募捐筹款。他曾回忆道，他的办学经验就是，每到一处就先去拜会当地的行政长官，说明来四川办学的目的，以寻求对方的支持。民国初年，毕启得到四川都督胡景伊以及陈延杰省长的支持，他们二位各自为大学捐出自己的俸禄3 000圆大洋，这是毕启为学校募集到的第一笔资金。

At the beginning of the university, from the school building, the purchase of teaching equipment to the appointment of teachers and so on all need huge expenditure. For a private university, all these expenses should be raised by itself, so president Beech had to raise money at home and abroad to keep the university running properly. He once recalled that every time he went to a new place, he visited the local administrator and explained his visiting parpose, in order to seek support. In the early years of ROC, Beech received the kindness of Sichuan Military Governor Hu Jingyi and Governor Chen Yanjie. The two each donated their own salary of 3,000 silver dollars to the university, which was the first fund Beech raised for the university.

随着一张张校园建筑设计图纸从荣杜易的笔下被设计出来，建造这些楼宇也就提上了议事日程。然而，修建这么多的大楼确实需要一笔不小的费用，才成立的大学经费捉襟见肘，校长毕启决定亲自前往美国募集建楼资金。

As soon as the campus architectural design drawings by Mr. Rowntree came out, the construction of these buildings was on the agenda. However, to build so many buildings did require a lot of money. The funds of just established university were tight. President Beech decided to go to the United States to raise funds.

临行前，校董事会的美国人周忠信告诉毕启，纽约有一位叫亚克门·柯里斯的医生是慈善家，柯里斯的慈善事业遍布全世界很多国家和地区，如果他肯为大学捐款，那数目一定不会少。

Before leaving, Joseph Taylor, an American on the university board, told Dr. Beech that a New York doctor named Dr. J. Ackerman Coles, a philanthropist, had charity in many countries and regions around the world. If he would donate money to the university, that amount must not be small.

毕启回到美国后，首先到纽约找到柯里斯，提出需要他为大学修建的楼宇提供资金的帮助。当时毕启给柯里斯看了正在修建的两栋学舍大楼的照片，柯里斯对其中一栋带有中国塔楼的学舍很感兴趣，说他可以付建设这栋学舍的费用。毕启又给他看了图书馆的设计图，但他对此毫无兴趣，他只对有塔楼的学舍情有独钟。他说："我就想要出资修建这栋学舍。"然而毕启告诉他已经有人出钱修建了这栋学舍时，他回答道："这有什么关系吗？叫他们退出来就行了，我来付款。"两个月后，捐赠人同意退出捐赠，柯里斯如愿以偿地捐赠了这栋学舍。他给毕启展示了一个钢雕，上面刻有他母亲名字"亚克门"的塔楼以及塔楼所在处，还刻有他如何获得这栋塔楼的故事。他告诉毕启，他想用这栋学舍纪念他的母亲，于是该学舍被命名为亚克门纪念室。之后，柯里斯又追加了10 000美元用作这栋学舍的维修费用。

When Dr. Beech returned to the United States, he first went to New York to meet Dr. Coles and asked him to provide financial help for the construction of the university buildings. He showed Dr. Coles pictures of the two university buildings under construction, and Dr. Coles was fascinated by one with a Chinese tower, saying he could pay for the building. Then Dr. Beech showed him plans of a library, but he had no interest in it, and he showed special favor to the tower building. "I want to pay for that building." Dr. Coles said. "But it is already paid." Dr. Beech replied. "It makes no difference" was his answer, "Get the donors off and let me get in." Two months later, Dr. Coles had the former donors agreeing. He donated the cost of the tower building. And he showed Dr. Beech a steel carving of the Ackerman Tower Building and the story of how he gained this tower. He also told Dr. Beech he wanted to name this tower Ackerman to memory his mother. Dr. Coles then added an additional $10,000 for the repairs to the premises.

亚克门纪念室外形为L形，转角处镶嵌一座中式八角重檐攒尖顶塔楼，该楼在20世纪90年代末被拆除。图片来源：四川大学档案馆。
The Ackerman Memorial Dormitory is L-shaped, and in the corner there is Chinese octagonal heavy eaves tower. The building was demolished in the late 1990s. Photo source: Sichuan University Archives.

后来，柯里斯又给毕启展示了他在纽约的家产，并对毕启说："有了这些，你们的大学将会永远地存在。"最终柯里斯把他家产的三分之一捐给了华西协合大学作为所有建筑物的永久维修费用。

Later, Dr. Coles showed Dr. Beech his property in New York and told Dr. Beech, "With these, your university will exist forever." Eventually, Dr. Coles donated one third of his property to West China Union University as costs for permanent maintenance of all the buildings.

亚克门纪念室是华西坝上最早修建的两栋大楼之一，该学舍外形为L形，有两层，在转角处镶嵌一座5层的中式八角重

1914年修建的亚克门纪念室，由美国柯里斯先生为纪念他的母亲亚克门而捐建。图片来源：1921年《华西协合大学校》。
The Ackerman Memorial Dormitory was built in 1914. It was donated by Dr. Coles, USA, in honor of his mother Ackerman. Photo source: 1921 West China Union University.

檐攒尖顶塔楼，塔楼的1~3层完全镶嵌在楼里，因而只有四面露出来，4~5层则完全高出整个屋顶，可以看见八个面。除了第二层的屋檐因与整座大楼的屋檐相吻合而不是飞檐外，其余的都是飞檐。从远处看去，塔楼就像是从一栋楼里长出来的，层层叠叠的飞檐更是营造出壮观的气势和呈现出中国古建筑特有的韵味。难怪柯里斯一眼就看上了这栋很有中国特色的建筑物，愿意捐助这栋学舍的修建费用。尽管亚克门纪念室看上去很奇特，就像两个不同类型的建筑物靠在一起，但这栋楼应该是四川最早的中西结合的建筑物。遗憾的是这栋楼在20世纪90年代末期被拆除了。

The Ackerman Memorial Dormitory is one of the first two buildings in Huaxiba. It is L-shaped, with two floors, and in the corner there is an inlaid 5-story Chinese octagonal heavy eaves tower. 1 to 3 floors of the tower are completely inlaid in the building, so only four sides exposed, and 4 to 5 of the tower are above the whole roof, eight sides can be seen. Except that the eaves on the second floor fit with the eaves of the entire building, the rest are all flying eaves. From a distance, the tower looks like "growing out" from the building. The overlapping eaves create a spectacular momentum and the unique charm of Chinese ancient architecture. No wonder Dr. Coles took a quick look at the picture of the Chinese building and was willing to donate to the cost of the building. Although the Ackerman Memorial Dormitory looks peculiar like two different types of buildings leaning together, the building should be the earliest building in the style of combination of Chinese and Western architecture in Sichuan. Sadly, the building was demolished in the late 1990s.

在毕启任校长的10多年间，他15次横渡大西洋和太平洋，穿梭于欧洲与美洲大陆，募集43万余美金用于购买校址、建筑校舍、添加教学设备等，建造了钟楼、事务所、图书馆、教育学院、医牙学院、理学院、生物楼等10多幢大楼。这些建筑都是捐赠者资助修建的，并以他们的名字命名。学校占地千余亩，校园里种植的2 000多株植物与中西合璧的建筑形成了独特的风景，而钟楼荷塘、鸳鸯小桥、对牛弹琴、柳塘压雪等景观使华西坝上的风光静谧而充满活力，生机勃勃却又不喧嚣。

During more than ten years, President Beech had crossed Atlantic Ocean and Pacific Ocean fifteen times, to and fro between Europe and the USA, raised more than $430,000 to purchase the university sites, build university buildings, add teaching equipment and so on. They built the clock tower, office, library, education college, medical and dental college, science college, biology building and other buildings. The buildings were funded by donors and named after them. The university covered an area of more than 1,000 mu, more than 2,000 plants were planted in the campus and the buildings of combination of Chinese style and Western style, all of which formed a unique scenery. The Clock Tower and Lotus Pond, Yuanyang Bridge (Lover Bridge), Cast pearls before swine, Snow pressing in liutang (lots of willow catkin on the playground) and along with other sight scenes, made Huaxiba a place of both tranquil and vitality.

华西坝老建筑的前世今生
The Stories of the Historic Architecture of Huaxiba

叁

走进华西坝

校门

Walk into the University Gate of Huaxiba

大学校门是学校与外界联系、进出的通道，是校园与周边区域的界线。作为城市的重要组成部分，校园是城市景观中独一无二的风景线，而校门则是学校形象的主要展示窗口，校门的格局是学校办学理念、品牌和地域文化的呈现。

The university gate is the connection and an access between the school and the outside world, and part of the demarcation line of the campus and the surrounding areas. As an important part of the city, the campus is a unique scenery line in the urban landscape, and the university gate is the main display window of the university image, and the presentation of the school-running philosophy, and regional culture.

设计图里的三种校门
The Three Gates in the Blueprints

从英国建筑师荣杜易的后代赵安祝提供给笔者的他的高祖父当年设计华西坝的建筑图纸判断，当时有三种校门的设计图纸，其中一种是中西合璧式校门，该校门体量宏大，犹如一座城池的城门。通过荣杜易绘制的效果图（里面绘有行人），目测该校门有30多米高，以中间拱券式门洞为主门，主门近10米高、5米宽，两旁的门洞要小得多。门楼为两层歇山式屋顶，在最上面的屋顶中间设计了一座方形钟塔。

Dr. Andrew George, a descendant of the British architect Fred Rowntree, provided to the author architectural drawings of the Huaxiba that his great-great-grandfather designed and drew in that year. There are three kinds of school gate design drawings. One drawing is a combination university gate of Chinese elements and Western elements. The gate is grand, like the gate of a city. Seeing from the effect drawing (with pedestrians) by Rowntree, visually measured the gate is about 30 meters high. The middle arch door is the main door, nearly 10 meters high, 5 meters wide, the door holes on both sides are much smaller. The gatehouse is a two-story gable and hip roof, with a clock tower built in the middle of the top roof.

荣杜易设计的华西协合大学校园鸟瞰图里的校门与前面提到的校门相比，更有明显的地域文化特点。校门位于校园中轴线的北端，面向城区，是一座中国传统的三间四柱牌坊式门楼，门楼采用歇山式屋顶，中间为拱券门，两旁建有八字影壁，对面建有一扇形的过街影壁。

上图：荣杜易手绘的华西协合大学校门。赵安祝提供。

The university gate manuscript. Hand-painted by Mr. Rowntree. Andrew George provided.

下图：荣杜易设计的华西协合大学校园鸟瞰图局部。赵安祝提供。

A bird's-eye view of the campus of West China Union University (part) designed by Mr. Rowntree. Andrew George provided.

In the bird's-eye view of the university campus designed by Mr. Rowntree, the gate, compared to the gate mentioned above has obvious characteristics of Chinese culture. The gate is located at the northern end of the central axis of the campus, facing the city. It is a traditional Chinese three room four-pillar archway gatehouse with saddle roof. The middle door is arch door. There are splayed door screen walls built on both sides of the gate. Opposite is a fan-shaped screen wall.

荣杜易设计的第三种校门为一座两柱牌坊门楼，比鸟瞰图里的校门形制和规模更小。

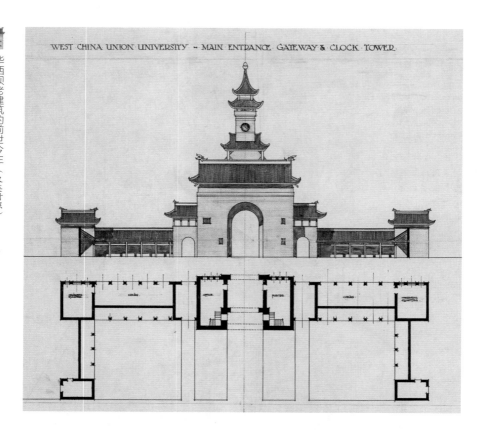

最终在营造校园的过程中，由于土地购买遇阻以及其他因素，校园大门的修建既没有采用荣杜易设计的宏大的中西合璧带钟塔的校门，也没有按照鸟瞰图里的设计位置与形式来修建。学校先后修建了两座有校名标志的校门，一座位于大学路的西边，临近大学事务所，便于与市区交流；另一座位于锦江边，便于使用水上交通，也就是荣杜易设计的第三种校门。

The third type of university gate designed by Mr. Rowntree is a two-pillar archway gate. It is smaller than the gate in a bird's-eye view of the campus.

Finally, in the process of building the campus, due to the obstruction of land purchase and other factors, the construction of the campus gate neither adopted the combined Chinese and Western style school gate with clock tower, nor was it built according to the design position and form in the bird's-eye view. The university has successively built two university gates with school name signs. One is located in the west of the Daxue Road, near the university administration building convenient to communicate with the city. The other is located by the Jinjiang River, convenient for water transportation, which is the third school gate designed by Rowntree.

荣杜易设计的华西协合大学校门正面图手稿。赵安祝提供。

West China Union University main gate front view manuscript designed by Mr. Rowntree. Andrew George provided.

从锦江边迁到人民南路上的校门
The University Gate Moving from the Bank of Jinjiang River to the South Renmin Road

1915年，荣杜易设计的一座两柱牌坊式校门在锦江边落成。该校门坐南朝北，面向锦江和城区。这座中式牌坊门楼大约9米高，其上方嵌有一方形红砂石牌匾，其上雕刻有"华西协合大学校 1910"字样，其背面嵌有一方形红砂石校牌，其上雕刻有"WEST CHINA UNION UNIVERSITY A.D.1910"字样。

In 1915, by the Jinjiang River, a two-pillar archway gate designed by Mr. Rowntree was completed, facing north, facing the Jinjiang River and the

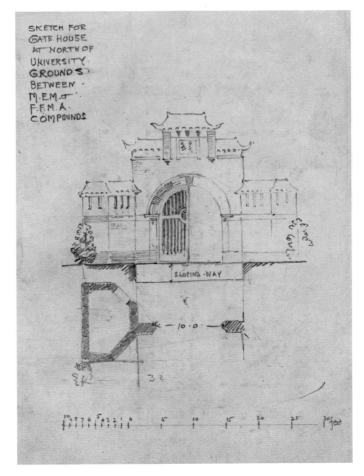

荣杜易手绘的大学校门草稿。赵安祝提供。
The manuscript of the drawings of the university gate, hand-painted by Mr. Rowntree. Andrew George provided.

从锦江边迁到人烟稀少的南坝上的校门
The University Gate Moving from a Riverbank of Jinjiang River to the South

从校门外向校园里拍摄的校门的照片，刻有"华西协合大学校1910"校名字样的红砂石匾牌镶嵌在牌坊式校门上。图片来源：四川大学档案馆。

Photos of the university gate taken at the out side of gate. A red sandstone plaque engraved with the university name "WEST CHINA UNION UNIVERSITY A.D.1910" in Chinese is inlaid on the gate. Photo source: Sichuan University Archives.

city. This Chinese archway gate is about 9 meters high. Above it there is a square red sand stone plaques on each side. The inside one engraved with the phrase "WEST CHINA UNION UNIVERSITY A.D.1910" on it, and the outside one engraved with the similar meaning characters in Chinese.

从荣杜易手绘草稿的批注可以知道，位于大学北面的这座校门原是为大学里美以美会和贵格会两个差会区域之间而设计的。后来在修建这座大门时在门上的前后两面有分别嵌有华西协合大学中文、英文校名的两块石碑，因而此门就成了华西协合大学的校门。

The notes on Rowntree's hand-drawn sketch tells that the gate is designed to be located in the north of the university grounds between M. E. M & F. F. M. A. area of the university. Later, when the gate was built, there were two stone tablets with the Chinese and English names

of the West China Union University embedded in the front and back sides above the door, so the gate became the gate of the West China Union University.

1939年6月11日，傍晚7点过，27架日本轰炸机飞来轰炸成都，华西坝五大学战时国际救护队队员黄孝逴（华西协合大学理学院制药系二年级学生）与同学正在学校附近一家餐馆吃晚饭，听到警报声她们立刻放下碗筷，离开餐馆返回学校到救护队集合地点集结。这时日本轰炸机的炸弹已经投下了，有四枚炸弹落在了华西坝上，就在黄孝逴她们快要进校门时，一枚炸弹落在校门附近，一枚弹片飞来击中了黄孝逴的后脑，她当场身亡。

1915年落成的中式华西协合大学校门，该牌坊式校门位于锦江边，从校门内向外看就可以看到河对岸的城墙。刻有"WEST CHINA UNION UNIVERSITY A.D.1910"校名字样的红砂石牌匾镶嵌在校门上。赵安祝提供。

The university gate completed in 1915 located by the Jinjiang River. The city wall on the other side of the river can be seen from the inside of the gate. A red sand and stone plaque with the university name "WEST CHINA UNION UNIVERSITY A.D.1910" is inlaid on the gatehouse. Andrew George provided.

On June 11,1939, after 7 p.m., 27 Japanese bombers came to bomb Chengdu. A member of Wartime International Rescue Team of Huaxiba Associated Universities in Chengtu, Ms. Huang Xiaochuo(a sophomore pharmacy school student) and classmates were having dinner at a restaurant near the university. When they heard the alarm, they immediately put down their chopsticks and bowls and left the restaurant to return to the university to the rescue team assembly site. By this time the Japanese bombs had been dropped, four bombs fell in Huaxiba. When Ms. Huang Xiaochuo and her classmates were about to enter the university gate, a bomb fell near the gate, and a shrapnel hit Ms. Huang Xiaochuo in the back of her head, she died on the spot.

消息传到重庆后，国民政府教育部通令嘉奖黄孝逴："该生奋勇捐躯，殊甚嘉许，自应特予褒扬。除由本部通令各校广为表彰，以昭激劝外，兹特发给国币五百元，即由该校立碑纪念。"

After the grievous news came to Chongqing, the Ministry of Education of the National Government praised Ms. Huang Xiaochuo. "She sacrificed her life, and she should be specially praised. In addition to the general recognition by the schools and encourage, special appropriation national currency 500 yuan was granted to the university to set up monument."

随着成都市政建设的发展，该门在1954年被拆毁，两块刻有校名的石碑也不知去向。1996年3月28日，在锦江边施工的工地上（该校门附近）出土了刻有华西协合大学英文校名的石碑。据了解，当年拆校门时这块刻有英文校名的石碑被学校员工用来盖水井，这样一来，这块石碑反而被无意间保留下来了。于是，这块消失了将近半个世纪的石碑终于重见天日。现在这块英文校牌石碑收藏在四川大学校史展览馆里。

With the development of Chengdu municipal construction, the gate was demolished in 1954, and the two stone plaques of university name were also missing. On March 28,1996, the stone plaque inscribed with West China Union University was unearthed at the bank of Jinjiang River construction site near the university gate. It is said in those years the stone plaque inscribed with the university name in English was covered on a well by someone, so that the stone plaque was inadvertently retained. This stone plaque disappeared for nearly half a century, finally appeared. And now it is collected in the University History Exhibition Hall of Sichuan University.

沧海桑田，位于南河边的校门已经不是学校与外界交流的主要通道。而当年校门被拆毁后，成都进行市政建设，以市中

2010年，四川大学华西校区校门重建，在原址上仿建了原华西协合大学中式牌坊门楼作为校门装饰。戚亚男摄于2021年。
In 2010, Huaxi Campus gate of Sichuan University was reconstructed, and the original Chinese style archway gate was imitated to build as the school gate decoration. Photo by Qi Yanan in 2021.

心原明代蜀王府（又称为皇城）为城市中轴点，修建了一条南北贯通的道路，总称人民路。学校位于人民南路三段，因修路而被划分成了东、西两个校区。于是学校在人民南路三段中间修建了东、西校区的两座校门，其形制为中式三间四柱大门，这样既方便师生往来于两校区，又方便校园与城市交流。东校门基本上替代了原锦江边的校门的功能，被大家视为学校的正门，自此，学校的校名牌都挂于此门上。20世纪50年代初，全国院系调整，学校改名为四川医学院。

　　Time brings great changes. Now the university gate located at the Jinjiang River bank is no longer the main passageway for communication with the outside world. For when it was demolished, Chengdu has carried out municipal construction, taking the former Shu Palace of the Ming Dynasty, also known as the imperial city as the central axis point of the city, a road running between north and south were built. It is called Renmin Road. This section of the university is the third section of South Renmin Road. The university is divided into east and west two campuses by South Renmin Road. Two school gates on the east and west campuses were built at the middle of the third section of South Renmin Road, so that this is convenient for teachers and students to communicate between the two campuses, and to communicate with the city. The gates shape is very simple Chinese three room four-column gate. The east gate basically replaced the function of the former university gate beside the Jinjiang River, regarded as the main entrance of the university. Since then, the university name plaque has been hung on this gate. In the early 1950s, the university was renamed to Sichuan Medical College.

1985年，为了发挥学校在对外交往方面的优势，加强国际、校际交流与合作，经卫生部批准，四川医学院更名为华西医科大学。

In 1985, in order to give full play to advantages of the university in foreign communication and strengthen international and inter-school exchanges and cooperation, approved by the Ministry of Health, Sichuan Medical College was renamed West China University of Medical Science.

2000年，四川大学（之前已与原成都科技大学合并）与华西医科大学合并，华西医科大学更名为四川大学华西医学中心。

In 2000, Sichuan University (previously merged with Chengdu University of Science and Technology) and West China University of Medical Science merged. Then, West China University of Medical Science is renamed West China Medical Center of Sichuan University.

2010年，人民南路改建，20世纪50年代修建在人民南路上的东校门被重建，在原址上仿建了原华西协合大学最早建在锦江边上的中式牌坊门楼作为校门装饰。消失了近半个世纪荣杜易设计的校门，又重新回到了人们的视野中。

In 2010, South Renmin Road was rebuilt, and the university east gate built on South Renmin Road in the 1950s was also rebuilt. At here an imitation gate was built to decorate the new university east gate. The imitation gate based on the Chinese style archway gate by the Jinjiang River. So, the university gate designed by Mr. Rowntree, disappeared for nearly half a century, was back to view.

消失了的大学路上的校门
The Disappeared University Gate on Daxue Road

1924年，荣杜易发表在英国杂志《建筑师》上的文章《华西协合大学》中有一张大学大门的照片。这张照片显示的是一座具有中国建筑风格的大门。大门高约3米，位于大学路西端。在大门内，最近的大楼是行政大楼。大门外是从东到西的街道。在这张照片的大门上还没有大学名字的牌匾。1939年美国人德慕克拍摄的

大学路上的校门挂上了写有"华西协合大学"校名的校牌。德慕克摄于1939年，美国卡尔顿学院提供。

The gate on the Daxue Road is hung the plaque of "West China Union University", taken by Paul Clifford Domke in 1939, Carlton College USA provided.

一张照片上显示，当时门上挂着一块"华西协合大学"的校名牌匾，而且门两边的耳室都加宽了。

In 1924, there was a photo of the university gate in the article *West China Union University*, which Mr. Rowntree published in the British magazine *The Builder*. The picture shows a Chinese house-style gate. The gate is approximately over 3 meters high, located at the western end of Daxue Road. Inside the gate the nearest building is the administration building. Outside the gate is street from east to west. There was not yet university name plaque on the gate in this photo. In 1939 an American Paul Clifford Domke took a photo shows there is a "West China Union University" name plaque hang on the gate, and the ear rooms on both sides were widened.

该校门出门左拐就是小天竺街，有不少居民以服务于大学为生。街上的小商铺沿街而开，比如裁缝店、皮鞋店、茶铺、理发店，甚至还有面包店和"Tip Top"西餐厅。抗战时期，南京的金陵大学、金陵女子文理学院，山东的齐鲁大学等学校内迁到华西坝与华西协合大学联合办学，一时间华西坝上云集了来自四面八方的学生和老师，这么多人的到来极大地刺激了小天竺街上的商业发展。比如学生多了，学校图书馆和教室的座位有限，许多学生课余就只能去茶铺看书，或者去休闲，因而这条街上茶铺的生意非常好。

Outside the gate, the left side is Xiaotianzhu Street. Many residents make a living by serving the university. So, many small shops on the street opened along the street, such as tailor shop, leather shoes shop, tea house, barbershop, even bakery and "Tip Top" West Restaurant. During the Counter-Japanese War, University of Nanking, Ginling College and Cheeloo University and others were moved to Huaxiba, jointly ran schools with West China Union University. Soon, students and teachers gathered from all directions in Huaxiba, and the arrival of so many people greatly stimulated the commercial development on Xiaotianzhu Street. For example, the students were much more, but the sitting space is limited in the university libraries and classrooms. Many students had to go to the tea house to read and relax after school. So the tea house business on the street was very good.

1939年1月13日下午，华西坝五大学学生战时服务团（简称战时服务团）邀请冯玉祥将军来华西坝作"坚持抗战到底"的演讲，当天冯玉祥就从这座校门进入学校作演讲。在怀德堂楼前，冯玉祥面对2 000多名来自中央大学、齐鲁大学、金陵大学、金陵女子文理学院和华西协合大学的学生作了非常激情的演讲，他说："纵观中国的抗战前途，是没有丝毫悲观……都知欺负我们老百姓的是日本，应当打日本打到底……确信最后的胜利是我们的，要大家努力，不要袖手不动。"

On the afternoon of January 13th,1939, the Student Wartime Service Group of Huaxiba Associated Universities in Chengtu invited General Feng Yuxiang to Huaxiba to make a speech of *Insist on Counter-Japanese War to the End*. That day, Feng Yuxiang entered the university though this gate for the lecture. In front of the Whiting Memorial Administration Building, Feng Yuxiang made a very passionate presentation to more than 2,000 students from National Central University, Cheeloo University, University of Nanking, Ginling College and West China Union University. He said, "Looking out the future of China's Counter-Japanese War, there is no slightest pessimistic... It is all known that Japan bullies our common people. We should fight against Japan to the end. Be sure that the final victory is ours! We should work hard, don't stand by."

演讲后冯玉祥又挥毫题写了"还我河山"几个大字留念，战时服务团赠送给冯将军一面锦旗，上书"深入民间"。

After the speech, Feng Yuxiang wrote "Restore Our Lost Territories" as a souvenir with brush pen. The Wartime Service Group gave General Feng a brocade flag with "Go Deep into the People" on it.

叁·走进华西坝——校门

这是从距离校门约30米的怀德堂北面楼上拍摄的照片。拍摄时间为20世纪20年代末到30年代初，校门位于右边，有两人刚刚进入校园。出校门的左边为小天竺街，街上有一路人，右边是大学路。校门对面是广益大学舍，远处有塔楼的建筑是1914年修建的亚克门纪念室。小天竺街早在清乾隆年间就有了。1910年华西协合大学开办后，广益大学舍和校园之间这段路1937年在学校的地图上标注为大学街，之后被称为大学路。校园的围墙很有特色，围墙不仅低矮，而且采用了中国园林砖墙的花式砌法，让校园的中西合璧楼宇等风景被路人尽收眼底。图片来源：四川大学档案馆。

The photo was taken in the late 1920s to early 1930s from the north upstairs of the Whiting Memorial Administration Building Hall, about 30 meters from the university gate. The university gate is located on the right, and two people have just entered the campus. On the left of the university gate is Xiaotianzhu Street, and there is a passer-by on the street. And on the right is Daxue Road. Opposite the university gate is the Friends' College Building. The distant building with a tower is the Ackerman Memorial Dormitory built in 1914. Xiaotianzhu Street has existed since the reign of Emperor Qianlong of the Qing Dynasty. After the West China Union University opened in 1910, the section of the road between the Friends' College Building and the campus was marked as Daxue Street on the map of the university in 1937, and was later known as Daxue Road. The wall of the campus is very distinctive. The wall is not only low, but also adopts the fancy masonry method of Chinese garden brick wall, so that the combination of Chinese and Western styles architecture and other scenery on the campus can be seen by passers-by. Photo Source: Sichuan University Archives.

20世纪50年代初院系调整后，学校更名为四川医学院，大学路上的校门上就挂上了四川医学院的校牌。

In the early 1950s, the university was renamed Sichuan Medical College, and the plaque of Sichuan Medical College was hung on university gate on Daxue Road.

自从四川医学院在人民南路三段上修建了新的校门以后，老校门就没有使用了。老校门上的华西协合大学校牌收藏在四川大学博物馆里，四川医学院的校牌现在已不知下落。20世纪70年代末，大学路成了自由农贸市场，老校门曾作为理发店使用。

This gate has not been used since the university built a new gate on the third section of South Renmin Road. The plaque of West China Union University on this gate was collected as cultural relics by Sichuan University Museum. The plaque of Sichuan Medical College is now missing. Daxue Road

校庆的日子里，大学路上的校门被装饰得喜庆而隆重，莘莘学子纷纷在此留影。邱蔚六提供。

On the anniversary of the university, the university gate on Daxue Road was decorated happily and ceremoniously, and many students took pictures here. Qiu Weiliu provided.

上图：荣杜易设计的修建在大学路上的校门。赵安祝提供。
The gate of West China Union University on Daxue Road. Mr. Rowntree designed. Andrew George provided.

下图：20世纪70年代末，大学路成了自由农贸市场，老校门曾作为理发店使用。戚亚男摄于1999年。
Daxue Road became a free farmers' market in the late 1970s, and the gate was used as a barber shop. Photo by Qi Yanan in 1999.

became a free farmers' market in the late 1970s, and the gate was used as a barber shop.

2006年，大学路实施老成都人文风情特色文化街改造工程，学校的砖围墙被拆除，用低矮的栅栏取而代之，这样一来，华西坝上的老建筑就亮了出来，同时这座老校门也被拆除了。

In 2006, an Old Chengdu Cultural Style Characteristic Street project was implemented at Daxue Road. The brick walls of the university were torn down, instead with low fences, so that the old buildings in Huaxiba appear in front of the public. Of course this old gate was also removed.

2020年年初，华西坝片区历史文化街区改造工程开始实施。华西坝是成都"八街九坊十景"中"九坊"之一，"九坊"为锦里、皇城坝、华西坝、音乐坊、水井坊、望江坊、大慈坊、文殊坊和猛追湾。经过一年多的打造，焕然一新的展示华西坝文化的大学路落成。临街清一色的玻璃校园围墙，让美丽的校园和那一幢幢民国时期修建的中西合璧的楼宇尽显于路人眼前。

In the beginning of 2020, the reconstruction project of Huaxiba Historical and Cultural Block began. Huaxiba is one of the old Chengdu Cultural Style Characteristic Streets. The project is one of the "Eight Streets, Nine Fangs(districts) and Ten Scenes" in Chengdu. Nine Fangs are Jinli, Huangchengba, Huaxiba, Music Fang, Shuijing Fang, Wangjiang Fang, Daci Fang, Wenshu Fang and Mengzhuiwan. After more than a year of reconstruction, a completely different Daxue Road showed up. The clear glass campus wall facing the street, let the beautiful campus and the architectures built at the period of ROC all in front of passers-by.

2021年改造后的大学路，行人走在街上可以透过玻璃围墙看到校园里面中西合璧的楼宇等建筑。这张照片里的老建筑就是学校的行政楼，玻璃围墙左端是荣杜易设计的原华西协合大学在大学路上的校门原址。戚亚男摄于2021年。

After the renovation of Daxue Road in 2021, pedestrians on the street can see the combined Chinese and Western buildings and other buildings inside the glass campus walls. The old building in this picture is the administration building of the university. The left end of the glass wall is the original site of the original university gate on Daxue Road designed by Mr. Rowntree. Photo by Qi Yanan in 2021.

华西坝老建筑的前世今生
The Stories of the Historic Architecture of Huaxiba

肆

华西坝的标志

钟楼

The Coles Memorial Clock Tower,
the Landmark of Huaxiba

钟楼，1926年竣工，高30米，由美国纽约的外科医生亚克门·柯里斯先生捐建。它不但是华西坝建筑群中的一座标志性建筑，也是成都市的地标性建筑。钟楼的屋顶是中国传统建筑形式歇山顶，而且设计师还别出心裁地在歇山顶上又建造了一个尺寸较小的四角攒尖顶的方形塔，看上去瘦长的塔身和尖的塔顶还显示出西方哥特式建筑的特点。尽管看上去上下两部分建筑风格不太融洽，但是它不经意地表达了在同一座建筑上中西两种不同建筑风格的特征，正如当时西方文化渐渐传入成都的状况。

The Coles Memorial Clock Tower, completed in 1926, 30 meters high, was donated by Dr. J. Ackerman Coles, a surgeon in New York, USA. It is not only a landmark of Huaxiba, but also a landmark of Chengdu. The roof of the clock tower is a traditional Chinese architectural form, and the designer has creatively built a smaller square

左图：这是拍摄钟楼的最佳位置，能看见钟楼南面与西面，一座高耸云霄的钟楼矗立在树林中间，呈现出一派田园风光景色；近处一条小溪从钟楼旁边缓缓流过；远处是嘉德堂和懋德堂。图片来源：四川大学档案馆。
This is the best place to take pictures of the clock tower. You can see the south and west sides of the clock tower, a soaring clock tower in the middle of the woods, a small stream slowly running by the tower, and the Biology Building and the Library building in the distance. What an idyllic scene! Photo source: Sichuan University Archives.

右图：中国建筑设计师古平南把钟楼屋顶原来的歇山顶改为中国传统十字脊歇山顶，同时采用平座的手法，在四面修建了观景台，这样一来钟楼更具有中国建筑的特色。戚亚男摄于2008年。
The Chinese architect Gu Pingnan changed the original gable and hip roof clock tower roof into the most complex cross gable and hip roof. He used the method of flat pedestal and built observation platform on all sides, so that the clock tower has more characteristics of Chinese architecture. Photo by Qi Yanan in 2008.

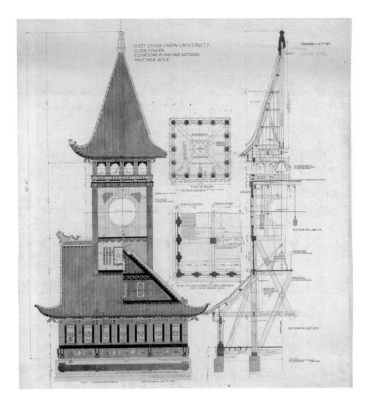

左图：荣杜易设计的钟楼正立面图、侧立面图、平面图和剖面图。赵安祝提供。

The clock tower front facade, side facade, plan and profile drawings. Mr. Rowntree designed. Andrew George provided.

右图：荣杜易设计的钟楼屋顶正立面图、侧立面图、平面图和剖面图。赵安祝提供。

The clock tower roof front facade, side facade, plan and profile drawings. Mr. Rowntree designed. Andrew George provided.

four-corner spire pyramidal on the top. The slender tower body and spire still show the characteristics of Western Gothic architecture. Although it seems that the upper and lower parts are not harmonious, it inadvertently expresses the characteristics of two different architectural styles of China and the West in the same building. Just as western culture was gradually introduced into Chengdu at that time.

1953年，在钟楼的维修过程中，中国建筑设计师古平南对塔顶做了很大的改动，把原来的歇山顶改为中国传统建筑屋顶形式中最为复杂的十字脊歇山顶，同时采用平座的手法，在四面修建了观景台，这样一来钟楼更具有中国传统建筑的特色，看上去也不别扭，上下两部分融为一体，形成一个更协调的整体建筑物。

During the maintenance of the clock tower in 1953, the Chinese architect Gu Pingnan made great changes to the top of the tower, changing the original gable and hip roof into the most complex cross gable and hip roof. At the same time, the observation platforms were built on four sides. Thus the clock tower is more characteristic of traditional Chinese

architecture, and it does not look awkward either. The upper and lower parts are integrated to form a more coordinated overall building.

正在建造中的钟楼，门洞清晰可见。赵安祝提供。
The clock tower under construction, the door hole is clearly visible. Andrew George provided.

　　华西坝钟楼的钟是由美国梅尼利制钟公司于1924年专门为华西坝钟楼制造的。大钟的心脏由两部分机械组成：一部分是走钟，用钟楼外面的四个钟面来显示时间；另一部分是打钟，用大锤敲击楼里的大钟来报时。走钟和打钟在时间上虽然同步，但各自的发条不同。走钟的发条需要每五天上一次，打钟的发条需要每三天上一次。该机械部分由美国塞思·托马斯制钟公司制造，从钟上的铭牌可以看到这是该公司制造的第2410台大钟机械，于1924年9月15日制成。

　　The clock in Huaxiba clock tower was specially manufactured for Huaxiba clock tower by Meneely Bell Co. Troy. N. Y. U.S.A. in 1924. The heart of the big clock consists of two mechanical parts. One part is the clock with the four clock faces outside the clock tower to display the time; the other part is the bell with a sledgehammer to strike the bell in the tower to tell the time. The clock and the bell are synchronized in time, but their clockworks are different. The clock needs to wind up every five days, and the bell needs to wind up every three days. The mechanical part is made by Seth Thomas Clock Co. U. S. It can be seen from the nameplate on the clock that it is the 2410th large clock machine produced by the company, and was made on September 15, 1924.

现在我们看到的大钟不像一般的钟那样悬挂着,而是固定在钟楼顶层地面上,由一柄大铁锤敲击钟的边沿而发声,大锤敲击的时间通过下面钟的机械装置来控制。这与荣杜易的设计是不同的,到底是当年在修建过程中就做了修改,还是在1953年维修中改变的,现在还没有证据证明是哪一种。

在钟的上面凸印着英文铭文:

梅尼利制钟公司
特洛伊 纽约
美国
公元1924年

戚亚男摄于2006年。

Now the bell is not suspended like the usual bell, but is fixed on the top ground of the clock tower, voiced by a sledgehammer striking on the edge of the bell. The striking time of the big hammer is controlled by the mechanical device of the clock below. This is different from Rowntree designed, and there is no evidence to prove whether it was modified during construction in 1926 or during repairs in 1953.The bell is embossed with English inscription:

MENEELY BELL CO.
TROY.N.Y
U.S.A.
A.D.1924

Photo by Qi Yanan in 2006.

钟楼里的大钟约有半人高,重380多千克。它不像一般的钟那样悬挂着,而是固定安装在顶层地面上,由一柄大铁锤敲击钟的边沿而发声。大锤敲击的时间通过下面钟的机械装置来控制。

The big bell in the clock tower is about half a person tall and weighs more than 380 kg. It is not suspended like the usual bell, but is fixed on the top ground, voiced by a sledgehammer striking on the edge of the bell. The striking time of the big hammer is controlled by the mechanical device of the clock below.

然而,随着时间的推移,长时间未维修的机械钟也走时不准了。从2006年起,学校停止了机械钟的运行,用电子钟替代了机械钟,钟声也使用录制的钟声播放。当然,原来的打钟被学校精心地保存于钟楼之内。

However, as time progressed, the mechanical clock had not been repaired for a long time, and was out of time. Since 2006, the mechanical clock has been replaced by an electronic clock, and the bell sound is also replaced by playing the recorded bell sound. Of course, the university carefully preserved the original machinery and bell in the tower.

2001年11月, 钟楼被成都市政府批准列为成都市首批文物建筑之一。

2002年12月,钟楼被四川省政府批准列为四川省第六批文物保护单位。

钟楼是华西坝的标志性建筑，每年毕业季毕业生都要穿上学位服到钟楼前拍照留念。钟楼见证了一届又一届的学生从华西坝毕业。咸亚男摄于2020年。

The clock tower is the landmark of Huaxiba. Every year during the graduation season in June, graduates wear their academic gowns and go to the clock tower to take photos. She has witnessed the graduating session one after one. Photo by Qi Yanan in 2020.

2013年5月，钟楼被国务院确定为全国重点文物保护单位。

In November, 2001, the clock tower was confirmed as one of the First Cultural Relic Buildings in Chengdu by the Chengdu Municipal Government.

In December, 2002, the clock tower was confirmed as the sixth batch of Cultural Relic Protection Units in Sichuan Province by the Sichuan Provincial Government.

In May, 2013, the clock tower was confirmed as a National Key Cultural Relic Protection Unit by the State Council.

钟楼从落成那天起就成为华西坝的标志性建筑，到今天它依然是。在近百年的岁月里围绕钟楼发生了许多故事，积淀了很深的文化内涵。

The clock tower has become the landmark of Huaxiba since its completion, and it is still so today. In the past hundred years, many stories have taken place around the clock tower, which accumulated deep cultural connotation.

城南钟声
The Bell Ringing in the South of City

早年间，钟楼地处城南边，四周除了大学的几栋建筑外就是农田和农舍，学生朗朗的读书声、田野里的家禽叫声伴随着钟声，给人一种世外桃源的感觉。每当钟楼的钟声敲响时，这浑厚的钟声传得很远，城里的市民也听这钟声知时辰。因而不少市民不称这钟声是华西坝钟声，而称其为城南钟声。

In the early years, the clock tower was located in the south of city. In addition to several university buildings, it was surrounded by farmland and farmhouses. The sound of students' reading and the sound of poultry in the fields accompanied by the bell gave people a feeling of a paradise. Whenever the bell rings, the thick bell ringing spreads far away, and the citizens of the city listen to the bell to know the time. Therefore, many citizens do not call this bell Huaxiba bell ringing, but the bell ringing of the city south.

民国年间，上海《申报》在刊登的《华西坝学界天堂》一文中介绍华西坝十景之钟楼时这样描写道："坝上的那座钟楼，它是全坝的计时之神，如巨人似的耸立在坝上，它年年月月，日日夜夜，忠勤地报告坝上人们的时辰。"

During the years of the ROC, the Shanghai newspaper *Shunpao* introduced the clock tower, one of ten views of Huaxiba. In the article *Huaxiba Academic Paradise*, It described this way, "the clock tower on the campus is the God of Timing of the whole campus. It stands on the campus like a giant. It faithfully tells the hours to the people on the campus year after year, day and night."

当年有人写华西坝："走出了成都的新南门，无数的茶馆立刻呈现在你的眼前，人声、琴声、歌唱声混在一起，使一个初来成都的人，顿时感觉到他们的悠闲和懒倦。由此向右转一个弯，便可以看见一座高耸云霄的钟塔，在傲然地矗立着，它睥睨着华西坝及其附近的人们，它是无上的权威者，控制着整个华西坝的时间和空间，再前进数十步，便到了现在成都的文化区了。"

In those days, someone wrote about Huaxiba, "After walking out of New South Gate of Chengdu, many tea houses immediately appeared in front of you. The voice of people, piano sound, and singing are mixed together, so that a newcomer to Chengdu can suddenly feels the leisure and slack of Chengdu people. From here, turn right, you can see a towering clock tower, standing proudly. It watches Huaxiba and the people around. It is an supreme authority, controlling the time and space of the whole Huaxiba. Take dozens of steps forward, you will reach the cultural area of Chengdu."

以后还有人用"起舞闻钟不用鸡"的词句来描写华西坝钟声的。城南的钟声可谓深入人心。

Later some one used the slang "The bell of Huaxiba replaced the cock crowing to tell the time" to describe the Huaxiba bell. The bell in the city south is deeply rooted in the hearts of the people in Chengdu.

成都第一家食品罐头厂的起源
The Origin of the First Food Cannery of Chengdu

成都第一家食品罐头厂是因维修华西坝钟楼的钟面而诞生的。1939年，华西坝钟楼的钟面坏了，需要换一个，当时钟面是用白铁皮做的。校方到城里的白铁铺里找铁匠来做这个直径近2米的钟面。一般人认为，用白铁皮做锅和蒸笼的师傅做一个圆形的面板应该是没有什么问题的，然而由于钟面太大，若不把这面板敲打得很平整，那是会影响钟针的走动的，因而不少白铁匠师傅都接不下这活路。最后，校方找到了当时业界有名的白铁匠徐师傅来做面板。徐师傅叫徐海亭，号鑫盛，白铁匠出身。徐师傅心灵手巧，在做面板的几天时间里，他结识了校方一位时常来看他做活的洋人，在他们闲聊时，谈到了饮食文化，洋人提到了外国的许多食品都做成罐头，便于保存食物，而且想吃时也很方便。说者无意，听者有心，徐师傅便动了自己做罐头的想法，他想到罐头容器是用白铁皮做的，而他自己就是干这一行的，敲个罐头盒子很容易，但要把食品做成罐头长时间保存那可要点学问。为此他在与洋人的交谈中特别仔细地问这问那，如罐头食品为什么能保持食物不变质、食物装进罐头后怎样抽气等。钟的面板做好后，徐师傅回家就开始琢磨做罐头的事，经过反复地试验，徐师傅终于摸索出了做罐头的工艺。随后徐师傅以自己的名号命名的鑫盛食品罐头厂开业了，从此在成都街面上就有了咱们成都造的罐头出售了。

The first food cannery of Chengdu originated from the maintenance of the clock face of the clock tower in Huaxiba. In 1939, the clock face of the clock tower was broken and needed to be replaced. At that time, the clock face was made of white iron sheet. The university sent people to the white iron shop in the city to find a blacksmith to make this clock face with a diameter of nearly two meters. Generally speaking, there should be no problem for a master who makes a pot and steamer with white iron sheet to make a round panel. However, because the clock face is too large, if the panel is not knocked flat, it will affect the movement of the clock needle, so many blacksmith masters can't take this job. Finally, they found the famous blacksmith Master Xu Haiting to do the panel. Master Xu, style name Xinsheng, had clever hands and good sense. During the few days of making the clock panel, there was a foreigner of the university often came to see Master Xu do his work. When they chatted, they talked about the food culture. The foreigner mentioned that many foreign foods are made into cans, which is convenient for preserving and eating. Master Xu had his idea of making a can. He thought that the canned container was made of white tin, and he was in this business himself. It is easy to knock a canned box, but the key knowledge of canning food and keeping them for a long time takes enormous time to learn. Therefore, in his conversation with foreigners, he asked many information carefully, such as why canned food can keep the food from deterioration, how to draw air out of the can after the food is canned, and so on. After the clock panel was done, Master Xu began to think about canning when he came home. After repeated experiments, Master Xu finally found out the process of canning. Then Master Xu opened Xinsheng Food Cannery named after his style name. From then on, we have cans made in Chengdu for sale on the streets of Chengdu.

钟楼前留影，送别投笔从戎的同学
A Special Picture Taken in Front of the Clock Tower to See Off the Student Who Has Joined the Army

在众多的华西学生与钟楼合影的老照片中，有一张照片是比较特别的。这张摄于1944年年底的照片上，几位学生爬上一棵并不粗壮的银杏树，以钟楼为背景照了这张合影照片。照片初看时给人的感觉是这几位学生咋这么顽皮，照合影照都要别出心裁地爬上树去照？但当你了解了照片背后的故事后，你肯定会有另外的感受。

Among many old photos of Huaxiba students taking photo with the clock tower, one photo is quite special. The photo, taken in late 1944, shows several students climbing a weak ginkgo tree and took this photo with the clock tower as the background. At first glance, it gives the impression that why these students are so naughty? Do they have to climb up the tree to take photos? But when you understand the story behind the photo, you will certainly have another feeling.

照片上的学生是1941年考入华西协合大学医牙学院的学生。入学时全班共有90多位同学，但经过3年的学习，只有13位同学幸运地没有被淘汰，留下来继续学习。没有被学校严格的淘汰制度所淘汰，本是很幸运的，谁不珍惜这个机会？可是国难当头，匹夫有责，当时许多热血青年纷纷应征入伍，奔赴前线抗日救国。许多热血沸腾的华西协合大学学子也投笔从戎，踊跃报名参军抗日。

The students in the photo were admitted to the Medicine and Dental College of West China Union University in 1941. There were more than 90 students in the class when entering the college. But after three years of study, only 13 students were lucky not to be eliminated and stayed on to continue their medical studies. This is very lucky, who does not cherish this opportunity? But every man is responsible for the national crisis. At that time,

11位医牙科同班同学在钟楼前送别叶成宗（前排右1）参加远征军合影留念。邓显昭提供。
Eleven classmates of Medical and Dental College took a photo in front of the clock tower to say goodbye to Ye Chengzong (front row right 1) who joined the Expeditionary Force. Deng Xianzhao provided.

many hot-blooded young people enlisted in the army and rushed to the front line to resist Japan and to save the country. Many students of West China Union University enthusiastically signed up to join the Counter-Japanese Army.

照片中就有这样一位来华西协合大学求学的泰国华侨，其名叫叶成宗，他自愿中断学业，报名参加远征军。即将出征印度的叶成宗与班上11位同学来到校园的钟楼旁，为了冲淡离别的惆怅气氛，他们攀上荷塘边的一株银杏树，大家都面带笑容尽量表达高兴的心情，留下了这张有华西坝标志——钟楼的纪念照。

In the photo, there is an overseas Chinese from Thailand, Ye Chengzong, who came to study at West China Union University. He voluntarily interrupted his studies and bravely participated in the expeditionary force and is about to go to India with the army. That day Ye Chengzong and his 11 classmates came to the clock tower on the campus. In order to dilute the melancholy atmosphere of parting, they climbed up a ginkgo tree beside the lotus pond, smiling and trying to express their happiness, leaving this memorial photo of the clock tower, the symbol of Huaxiba.

当晚同学们准备了一些食品在校园里郑重地给叶成宗同学举行了告别会。在告别会上，大家觉得叶成宗同学即将上抗日前线，生死未卜，他们这个集体少了一位朝夕相处的同学，依依不舍的离别之情使他们激动地决定：他们要像三国时桃园三结义那样，相互间以兄弟姐妹相称，即便叶成宗同学离开了这个集体，他也是他们的兄弟。随后，他们就以年龄的大小排序称呼：老大梁永耀、老二邓显昭、老三叶成宗、老四黄安华、老五胡镜尧、老六薛崇成、老七张慎微、老八温光楠、老九杨式之、老十刘嘉伦、老十一唐维晶、老十二郭媛珠、老十三刘源美。

That night, the students arranged some food and solemnly held a farewell party for Ye Chengzong on campus. At the farewell meeting, everyone felt that Ye Chengzong was going to the Counter-Japanese front, and his life and death were uncertain. Their collective would lack a classmate who lived together day and night. Their reluctant parting made them excitedly decide that they should be commensurate with each other as brothers and sisters, just like the three sworn brothers in Taoyuan during the Three Kingdoms period. Even if Ye Chengzong left the collective, he was also their brother. Then they called themselves in order by age: the eldest Liang Yongyao, the second Deng Xianzhao, the third Ye Chengzong, the fourth Huang Anhua, the fifth Hu Jingyao, the sixth Xue Chongcheng, the seventh Zhang Shenwei, the eighth Wen Guangnan, the ninth Yang Shizhi, the tenth Liu Jialun, the eleventh Tang Weijing, the twelveth Guo Yuanzhu, the thirteenth Liu Yuanmei.

此后叶成宗告别了同学，告别了老师，告别了华西坝，加入到抗击日寇的行列中。按照当时国民政府的规定，青年学生中断学业参军抗战，战争结束后可回校继续完成学业。所以抗战胜利后，1946年叶成宗又回到了华西坝继续学习，只是不可能再回到原来的年级了。

Then, Ye Chengzong said goodbye to his classmates, to his teachers, to Huaxiba and joined the ranks of fighting against the Japanese invaders. According to the regulations of the national government at that time, young students who interrupted their studies to join the Counter-Japanese War, after the war can return to school to continue their studies. Therefore, after the victory of the Counter-Japanese War, Ye Chengzong returned to Huaxiba to continue his study in 1946, but it was impossible to return to his original class.

华西坝钟楼的兄弟——安仁中学钟楼
Copies of the Huaxiba Clock Tower

四川省大邑县安仁镇的老公馆建筑群以中西合璧的风格在全国古镇建筑中独树一帜。2004年，一座高21米的钟楼修建在安仁镇镇口。这座具有华西坝钟楼外形特色的钟楼屹立在镇口欢迎着来安仁镇的客人们。它是改革开放以来安仁镇的标志性建筑。了解华西坝的历史、见过华西坝老建筑的人会感到诧异，怎么在这里会有一座与华西坝钟楼相类似的钟楼呢？当地人或是导游会告诉你，这是县里为了体现安仁古镇的古风古貌，特别仿安仁中学校园里的钟楼而建造的。难道安仁中学的钟楼与华西坝的钟楼很相像？那哪个在先？哪个模仿哪个呢？

The old residence complex in Anren Town, Dayi County, Sichuan Province is unique in the ancient town architecture with a combination of Chinese and Western style. In 2004, a 21-meter-high landscape clock tower was built at the entrance of Anren Town. This landscape clock tower with the appearance of Huaxiba clock tower stands at the town entrance to meet the guests to Anren town. It is the landmark building of Anren town since the Reform and Opening Up. People who understand the history of Huaxiba and have seen the old buildings of Huaxiba will be surprised. How is there still a clock tower similar to the Huaxiba clock tower here? Locals or tour guides will tell you that this was built in the county to reflect the ancient appearance of Anren ancient town, especially imitated the clock tower on the campus of Anren Middle School. Is the clock tower of Anren Middle School very similar to that of Huaxiba? Which one comes first? Which one imitated the other?

其实只要我们了解一下两座钟楼修建的时间，就可以知道安仁中学的钟楼是模仿华西坝的钟楼而修建的，因为华西坝的钟楼是在1925年修建的，而安仁中学的钟楼是1944年才修建。从外形上看两校的钟楼简直就是一对双胞胎兄弟，只是一大一小。华西坝的钟楼高30多米，下半部分为一歇山式屋顶的塔楼，最为特别的是设计师别出心裁地在歇山式屋顶上又建造了一个较小的四角攒尖屋顶塔楼，这种形式的塔楼在现在的华西坝建筑群里还能看到。而安仁中学的钟楼高近15米，下半部分为一歇山式屋顶的塔楼，上半部分也是一个四角攒尖屋顶的塔楼，只是其中间部分多了一个歇山式屋顶。因而对熟悉华西坝建筑的人来说，晃眼一看，安仁中学的钟楼给人的感觉是似曾相识，但又有点怪怪的。当然，现在华西坝的钟楼与安仁中学的钟楼又有不同。20世纪50年代初，华西坝钟楼经过维修，把下部分的歇山式屋顶改为了最为复杂的十字脊歇山顶，并在四周加建了观景台。

左图：建于1925年的华西坝钟楼。摄于1926年。这张照片是黑白照片上了彩色。赵安祝提供。

The Huaxiba clock tower, built in 1925. This color photo was originally black and white. Photo taken in 1926. Andrew George provided.

右图：建于1944年的大邑县安仁镇安仁中学钟楼。戚亚男摄于2007年。

The clock tower of Anren Middle School, built in 1944. Photo by Qi Yanan in 2007.

In fact, if we know the construction time of the two clock towers, we can know that the clock tower of Anren Middle School was built in imitation of the clock tower of Huaxiba, because the clock tower of Huaxiba was built in 1925, while the clock tower of Anren Middle School was built in 1944. From the appearance of the clock towers of the two schools, they look like twins, the only difference is size. The clock tower of Huaxiba is nearly 30 meters high, and the lower half is a tower with a saddle roof. The most special is the designer's ingenious construction of a smaller four-corner pointed roof tower on the saddle roof. Now this form of tower can still be seen in the Huaxiba complex. The clock tower of Anren Middle School is nearly 15 meters high, and the lower half is also a tower with a saddle roof. The upper half is also a four-corner tower, but at the middle of the upper part there is a saddle roof. Therefore, for those who are familiar with Huaxiba architecture, glancing at the clock tower of Anren middle school feel like they have met, but a little strange. Of course, now the clock tower of Huaxiba is more different from that of Anren Middle School. In the early 1950s, the clock tower of Huaxiba was repaired. The lower part original saddle roof was changed into the most complex cross saddle roof. At the same time, the observation platforms were built on four sides.

1996年，安仁中学的钟楼作为刘氏庄园的附属建筑，被国务院确认为国家重点文物保护单位。

In 1996, the clock tower of Anren Middle School, as a subsidiary building of Liu's Manor, was confirmed as a National Key Cultural Relic Protection Unit by the State Council.

高考作文《华西坝的钟声》
University Entrance Examination Composition
The Bell Ringing of Huaxiba

1946年7月24日,华西协合大学招生考试第一天,照例是考国文,有两道作文题目任选:一是《华西坝的钟声》(白话文);二是《学习国文之回顾》(文言文)。大多数考生选的是《华西坝的钟声》,因为华西坝的钟楼和钟声对他们来说是很熟悉的。就在考试的前几天,当地报纸还写道:"华西大学在四川私立的大学比较上还是一个好的学校,它有天然优美的环境,使人艳羡的钟鼎式的洋楼,设备完善的实验室,有好多青年的灵魂已经绕在它的四周。"

On July 24, 1946, the first day of the West China Union University admission examination, the Chinese essay had two topics optional: one was *The Bell Ringing of Huaxiba* (vernacular); the other was *Review of Learning Chinese* (classical Chinese). Most candidates chose *The Bell Ringing of Huaxiba*, because they were very familiar with Huaxiba clock tower and bells. Just a few

肆·华西坝的标志——钟楼

左图：钟楼春色。咸亚男摄于2020年。
Clock Tower in spring scenery. Photo by Qi Yanan in 2020.

中图：钟楼繁花似锦。咸亚男摄于2020年。
Flowers blossom around Clock Tower. Photo by Qi Yanan in 2020.

右图：钟楼秋色。咸亚男摄于2008年。
Clock Tower in autumn scenery. Photo by Qi Yanan in 2008.

days before the examination, a local newspaper wrote, "West China Union University is a competitive university among private universities in Sichuan. It has a natural and beautiful environment, beautiful Chinese and Western architectural style buildings, well-equipped laboratories. It is the yearning place of young people".

当年考生们笔下是如何描写华西坝的钟声我们现在已不得而知。不过好在当年有一位名落孙山的女生，第二年参加补习时专门请她的老师——一位前几年从华西协合大学毕业的高才生，也以"华西坝的钟声"为题，替她写范文供她参考。这位老师欣然命笔，写了近3 000字的文章，他的文采博得姑娘的称赞。很多年以后，这位老师都还记得他当时写的一些句子：

How did the candidates describe the bell ringing of Huaxiba? No one knows now. However, that year a girl failed to pass the examination. The next year she invited her teacher, an excellent student graduated from West China Union University a few years ago, to write a model essay under the title *The Bell Ringing of Huaxiba* for her reference. The teacher gladly wrote an article of nearly 3,000 words. His literary talent delighted the girl. Many years later, the teacher still remembered some sentences he wrote:

55

"华西坝的钟声，不同于深山古庙的，使人静心寡欲，超越人寰的钟声。它唤起人内心求知的渴望。"

"The bell ringing of Huaxiba is different from that of the ancient temple in the deep mountains, which makes people calm, have few desires, beyond the world. It evokes people's inner desire for knowledge."

"钟声响了，华西坝从梦里醒来，揉着惺忪的眼，望着一片灰蒙蒙的天，半月池的游鱼被荷花落瓣惊散。"

"When the bell rang, Huaxiba woke up from his dream. Rubbing his loose eyes and looking at the gray sky. The fishes in the half moon pool were frightened to scatter by the falling petals of the lotus."

"它报导着作息的时间，男女同学坐在窗明几净的教室听教师苦口婆心地讲课：知识是那么新鲜，道理是那么玄远。"

"The bell reports the time of work and rest. Male and female students sit in the classroom with clear windows and clean tables, listening to the teacher's earnest and diligently lectures: knowledge is so fresh and the truth is so profound."

"钟声响了，迎来西国朋友轩昂的气势，骄傲的步伐。环境优美的华西坝仿佛是他们开创的天堂，使他们乐不可支，忘记了脚下踏的是中国的土地。喧宾夺主有失礼貌仪态。"

"The bell rang, ushered in the western friends with proud pace of the majestic momentum. The beautiful environment of Huaxiba seems to be a paradise created by them, which makes them so overjoyed that they forget they are stepping on the land of China. It is bad manners presumptuous guests usurp the host's role."

华西坝老建筑的前世今生
The Stories of the Historic Architecture of Huaxiba

伍

华西校区行政楼

怀德堂

The Whiting Memorial Administration
Building of Huaxiba Campus

紧邻大学路西边的这栋大楼当时叫事务所，又名怀德堂，由美国纽约的恩甫夫妇捐建。大楼1915年动工，1919年建成。楼长53米，宽28米。原华西协合大学的行政办公室、礼堂、教师休息室、学生俱乐部、教室和摄影工作室都位于该楼内。现该楼为四川大学华西校区行政楼。

The building next to the west of Daxue Road is the Whiting Memorial Administration Building, donated by Mr. and Mrs. Joseph B. Morrell, New York, U.S.A. The construction began in 1915 and was completed in 1919. The building is 53 meters long and 28 meters wide. It used to be the administrative offices of West China Union University. And auditorium, faculty common room, students' club, class rooms and the photographic studio were also located in this building. Now it is the office building of Huaxi Campus of Sichuan University.

怀德堂为H形对称形式，前廊由8根粗圆柱采用具有四川地方建筑特色的穿斗式梁架结构支撑。屋顶主体为中式的重檐歇山顶，重檐下宽上窄，上下差别较大，而西式建筑特有的老虎窗、烟囱等造型也在屋顶上同时出现；侧面采用中国传统建筑的门楼结构，穿出屋面做成凸出的雉堞，还做成中式牌坊的形式。这样看上去侧面造型丰富，就像中国的庙宇、牌坊，具有中国传统建筑的味道。

1919年落成的怀德堂正面。图片来源：四川大学华西医学中心提供，由芳卫廉女儿赠。
The front of the Whiting Memorial Administration Building completed in 1919.
Photo source: West China Medical Center, Sichuan University provided, a gift from William P. Fenn's daughter (Mary Francis Hazeltine).

怀德堂正面。咸亚男2010年摄。
The front of the Whiting Memorial Administration Building. Photo by Qi Yanan in 2010.

The Whiting Memorial Administration Building is H-symmetrical form. The front porch is supported by eight thick columns and a column and tie frame structure with local architectural characteristics in Sichuan. The main body of the roof is the Chinese heavy eaves gable and hip roof. The obvious the heavy eaves is lower wide and upper narrow, which is largely difference in volume. The roof windows, fireplace chimneys and other shapes of western-style buildings also appear on the roof at the same time. The sides adopt the gatehouse structure of traditional Chinese buildings, through the roof to made into protruding battlements and also in the form of Chinese archway. All these make the side styling looks rich, just like Chinese temples and memorial archways, and have the style of traditional Chinese architecture.

让人想不到的是，荣杜易在设计怀德堂时不放过任何细节，如奠基石上标记该建筑的建造年代是用两种方式，一种是西式公元纪年——A.D.1915，另一种是民国纪年法——中华民国四年；室外台阶与室内楼梯扶手柱头上都是采用莲花座造型，可以看出荣杜易用在中国人广为信奉的佛教里有着重要地位的莲花座装饰建筑物的良苦用心；而更让人想不到的是，荣杜易在小小的瓦当上也下了功夫，他设计把"华西协合大学"与"民国四年"的字样烧制在瓦当上。

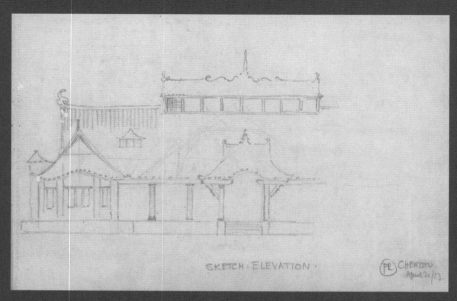

1913年4月23日，荣杜易在成都手绘的怀德堂草稿。赵安祝提供。
The sketch of the Whiting Memorial Administration Building hand-painted by Mr. Rowntree in Chengdu on April 23, 1913. Andrew George Provided.

荣杜易设计的怀德堂侧立面图。赵安祝提供。
The Whiting Memorial Administration Building side elevation diagram designed by Mr. Rowntree. Andrew George provided.

Surprisingly, Mr. Rowntree never let go of any details in the design of the Whiting Memorial Administration Building, and marked the construction date of the building on the foundation stone in two ways: one is the Western-style — A.D. 1915, and the other is the Chinese-style — the fourth year of the ROC. Outdoor steps and indoor stair handrail column heads are the shape of the lotus seat. It can be seen that Rowntree used the good intentions of decorating the buildings with the lotus seat, which has an important position in the widely faith Buddhism in China. And more unexpected is that Rowntree also made efforts on the eaves tile. He designed the words "West China Union University" and "the fourth year of the Republic of China" on the eaves tile.

1914年11月,从成都寄给荣杜易的怀德堂地基的施工现场照片。照片背面写道:"在壕沟里交替铺上鹅卵石和黏土,先用小夯锤捣四圈,然后四人用大夯锤捣四圈。用长杆夯实一个旧井的垃圾填充物,这是在锤打沟渠后才发现的,当时这个地方显得又软又湿。"
照片上除了用布裹在头上的正在施工的民工外,有两位外国人特别显眼,一位身穿中式服装,头戴鸭舌帽,脚穿圆口布鞋,手提公文包的外国人,像是工程的承包商,而另外一位西装革履打扮的外国人又好像是建筑工程师。赵安祝提供。

This is the photo of the construction site of the Whiting Memorial Administration Building foundation sent from Chengdu to Mr. Rowntree in UK in November, 1914. The back of the photo says, "The trenches are first hammered. Cobblestones are being laid in the trenches, alternating with layers of clay, which is first tamped four rounds with the small tampers, then four rounds with the large hammers by 4 men. Using the long pole to tamp down the rubbish fill of an old well that was discovered only after hammering the trench, when this place appeared soft and wet."
In the photo, in addition to the migrant workers under construction with cloth wrapped around their heads, two foreigners are particularly eye-catching. One is wearing Chinese clothes, hats, round mouth cloth shoes and a briefcase in his hand. He looks like a contractor. The other is wearing a suit and tie. He seems to be a construction engineer. Andrew George provided.

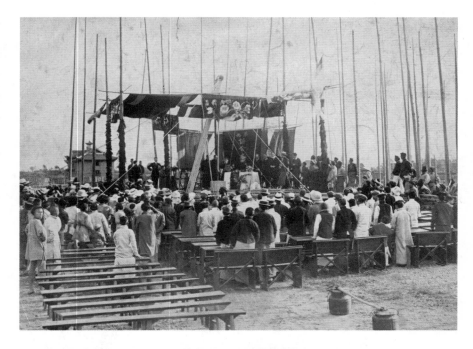

1915年怀德堂奠基仪式。赵安祝提供。

The foundation stone laying ceremony of the Whiting Memorial Administration Building in 1915. Andrew George provided.

荣杜易设计的怀德堂屋顶和门楼脊兽。赵安祝提供。

The Whiting Memorial Administration Building roof and ridge beast of gatehouse designed by Mr. Rowntree. Andrew George provided.

20世纪50年代，在维修怀德堂时对屋顶与门楼上的脊兽做了很大的改动。原来屋顶上的两条龙的尾巴朝向屋顶中间的宝顶，维修后改变为二龙背靠中间花篮，还增添了美观的水波浪；而门楼上的脊兽改为跳跃的两条鱼，中间的宝顶改为一个扇形装饰物。

During the maintenance of the Whiting Memorial Administration Building in the 1950s, great changes were made to the ridge beasts on the roof and the gatehouse. Original the tails of the two Loongs on the roof were towards the precious dome in the middle of the roof. After maintenance, two Loongs backs are against the middle flower basket, and around them added beautiful water waves. The ridge beast on the gatehouse was replaced by two jumping fish, and the precious dome in the middle was replaced by a fan-shape decoration.

2001年11月，怀德堂被成都市政府批准列为成都市首批文物建筑之一。
2002年12月，怀德堂被四川省政府批准列为四川省第六批文物保护单位。
2013年5月，怀德堂被国务院确定为全国重点文物保护单位。

In November, 2001, the Whiting Memorial Administration Building was confirmed as one of the First Cultural Relic Buildings in Chengdu by the Chengdu Municipal Government.

In December, 2002, the Whiting Memorial Administration Building was confirmed as the sixth batch of Cultural Relic Protection Units in Sichuan Province by the Sichuan Provincial Government.

In May, 2013, the Whiting Memorial Administration Building was confirmed as a National Key Cultural Relic Protection Unit by the State Council.

首任校长美国人毕启
American Beech, the First President of the University

怀德堂是学校行政事务的中心，校长的办公室就设在该楼。学校首任校长是美国人毕启，他1867年出生于英国，6岁时随其父母迁居美国。小学毕业后，毕启便打工赚取学费。1899年，毕启毕业于惠斯廉大学文学院获学士学位，第二年来到四川。毕启在重庆曾家岩办学堂5年后，他看到当时整个西南地区尚无现代高等学府。1905年，他联合英国、美国和加拿大三国的4个差会在中国西部重镇成都创办了一所综合性大学——华西协合大学。

The Whiting Memorial Administration Building is the center of the university's administrative affairs, and the office of the university president is located in the building. The first president was Beech. He was born in Britain in 1867 and moved to the United States with his parents at the age of 6. After graduating from primary school, Beech worked to earn tuition. In 1899, Beech graduated from the school of Arts of Wesleyan University with a bachelor's degree and came to Sichuan the next year. After five years of running a school in Zengjiayan, Chongqing, Dr.

Beech discovered that there was no modern institution of higher learning in the whole southwest at that time. In 1905, he united four missionary societies from Britain, the United States and Canada to jointly establish a comprehensive university in Chengdu, an important town in Western China. The university was West China Union University.

在毕启任职的10多年间，他15次横渡大西洋和太平洋，穿梭于欧洲与美洲大陆，募集43万余美元用于学校的建设，为学校的发展做出了巨大的贡献。

During more than ten years as the university president, Dr. Beech had crossed Atlantic Ocean and Pacific Ocean fifteen times, to and from Europe and the USA, raised more than $430,000 for the construction of the university, making great contributions to the development of the university.

首任华人校长张凌高
Zhang Linggao, the First Chinese President of the University

1933年9月，中华民国教育部同意华西协合大学立案，学校正式得到国民政府的认可。怀德堂迎来了首任华人校长张凌高，毕启校长退位任校务长。张凌高，1890年生于四川省璧山县（现为重庆市璧山区）。1919年，张凌高在华西协合大学毕

1924年，华西协合大学在中国西部地区开先河，招收了8位女生，实行男女合校。校长毕启在怀德堂迎接这8位女生进入高等学校读书学习。图片来源：四川大学档案馆。

In 1924, West China Union University opened a precedent in Western China, enrolled 8 girls and implemented education of male and female co-school. President Beech welcomed the eight girls into the university at the Whiting Memorial Administration Building. Photo source: Sichuan University Archives.

业，获文学学士学位，之后两次留学美国，获得硕士和博士学位。张凌高的走马上任是中国人收回教育主权的体现，从此学校在他的领导下历经艰辛成为西南地区有名的一所高等学府。

In September, 1933, the Ministry of Education of the ROC approved the registration of West China Union University, which was recognized by the national government. Zhang Linggao was the first Chinese to be the university president. He was welcomed in the Whiting Memorial Administration Building and Dr. Beech abdicated to be a director of the university affairs. Born in 1890 in Bishan County, Sichuan Province (it is now Bishan District of Chongqing), Zhang Linggao graduated from West China Union University with a bachelor of Arts degree in 1919, and then studied in the United States twice, obtained his master's and doctoral degrees. Dr. Zhang Linggao taking office is the embodiment of the Chinese people's recovery of educational sovereignty. Since then, under his leadership, the university has gone through hardships and become a famous institution of higher education in Southwest China.

抗战期间，张凌高带领全校师生接纳了先后迁到华西坝的南京中央大学医学院、金陵大学、金陵女子文理学院、济南齐鲁大学，苏州东吴大学生物系，北平燕京大学、北平协和医学院护士专科学校的部分师生。一时间华西坝上云集了数千名师生，教学设施有限，如何安排这么多学生上课是个难题。张凌高发挥他的才智，他倡议各大学走联合办学的道路，按师资的专长特点，充分利用教学资源，统一安排分工，各校分别开课，教师可以跨校讲课，学生可跨校选各校课程上课，学校承认所读学分。其间，五大学的校长和各系处定期召开联席会议，研究教学、科研、校务等问题。联合办学这段时期，不仅对成都的医疗卫生、高等教育及华西协合大学本身都有很大的促进和发展，而且为国家培养了大批人才。

During the Counter-Japanese War, Dr. Zhang Linggao led all the teachers and students to accept some teachers and students from Central University School of Medicine, University of Nanking, Ginling College, Cheeloo University, Department of Biology of Soochow University, Yenching University and Nurse College of Peking Union Medical College. Soon, thousands of teachers and students gathered in Huaxiba. With limited teaching facilities, it was extremely hard to supply enough classrooms for so many students. Dr.

华西协合大学首任华人校长张凌高。
图片来源：四川大学档案馆。
Dr. Zhang Linggao, the first Chinese president of West China Union University. Photo source: Sichuan University Archives.

抗战时期华西坝五大学校长联席会议在怀德堂里举行。从左起依次为燕京大学代理校长马鉴、金陵女子文理学院吴贻芳校长、金陵大学陈裕光校长、华西协合大学张凌高校长、齐鲁大学汤吉禾校长。图片来源：耶鲁大学。

During the Counter-Japanese War, the joint meeting of presidents of Huaxiba Associated Universities in Chengtu was held in the Whiting Memorial Administration Building. From the left, Ma Jian, acting president of Yenching University, Ginling College President Wu Yifang, University of Nanking President Chen Yuguang, West China Union University President Zhang Linggao, Cheeloo University president Tang Ji he (Edgar Chi-ho Tang). Photo source: Yale University.

Zhang Linggao played his sagacity. He proposed that the universities take the road of running schools jointly, make full use of teaching resources according to the specialty characteristics of teachers, and arrange the division of job in a unified way. Each school opens classes separately, teachers can give lectures across schools, students can choose courses across schools, and the school will admit the credits they have read. During the meantime, regular joint meetings of presidents and departments of the Associated Universities in Chengtu were established to study teaching, scientific research, school affairs and other issues. During this period, the joint education not only promoted and developed Chengdu's medical and health care, higher education and West China Union University itself, but also trained a large number of talents for the country.

　　1945年6月，华西坝又迎来了五大学联合毕业典礼。华西协合大学校长张凌高、金陵大学校长陈裕光、金陵女子文理学院校长吴贻芳、齐鲁大学校长刘世传和燕京大学代理校长马鉴，以及四川省政府主席张群、川康绥靖公署主任邓锡侯和四川省教育厅厅长郭有守等军政要员参加了这次毕业典礼。郭有守在典礼上讲道，二战将结束，中、美、英、苏、法五个强国对轴心国的战事已是胜利在望了。各校从敌占区千里迢迢内迁复校，历尽艰苦，精神可佩。来蓉后聚合而成坝上五大,造就不少人才,为国献力。你们五大学府就好比五大强国,团结一致,才能不断取得胜利!

In June, 1945, the joint graduation ceremony of the Associated Universities in Chengtu was held in Huaxiba. West China Union University President Zhang Linggao, University of Nanking President Chen Yuguang, Ginling College President Wu Yifang, Cheeloo University President Liu Shichuan, Yenching University Acting President Ma Jian, President of Sichuan Provincial Government Zhang Qun, Director of Chuankang Appeasement Department Deng Xihou, and Director of Sichuan Provincial Education Department Guo Youshou attended the graduation ceremony. Guo Youshou said at the ceremony, World War II will end, China, the United States, Britain, the Soviet Union, France, the five powers of the war against the Axis powers is in sight of victory. The schools moved thousands of miles from the enemy occupied areas to inland and resume schools, went through hardships. The spirit can be admired. After the universities came to Chengdu and jointed to be Associated Universities in Chengtu in Huaxiba. Associated Universities in Chengtu educated a lot of talents, contributed for the country. Your Associated Universities in Chengtu are just like five great powers. Only be united as one, can continue to win!

在怀德堂举行抗战宣传活动
Counter-Japanese War Activities in the Whiting Memorial Administration Building

怀德堂是华西协合大学的行政中心所在地,抗战期间国内外不少政要人员来华西坝都在此地留下了身影。最早来怀德堂宣传抗日的是冯玉祥将军。

The Whiting Memorial Administration Building is the center of West China Union University's administrative affairs. During the Counter-Japanese War, many dignitaries at home and abroad came to Huaxiba. The first one who came to the Whiting Hall to publicize the Counter-Japanese War was General Feng Yuxiang.

1939年1月2日,冯玉祥离开重庆到成都作抗日宣传演讲,沿途经永川、隆昌、内江、资阳,一路上都作演讲。13日下午4点,应华西坝五大学战时服务团之邀,冯玉祥一身戎装来到华西坝作演讲。原定在怀德堂二楼礼堂作演讲,但前来听演讲的学生有2 000多名,而礼堂只能容纳两三百人,只好把演讲地点改在怀德堂的门前,让冯将军站在一张方桌上面作演讲。

January 2, 1939, Feng Yuxiang went from Chongqing to Chengdu to make a speech on Counter-Japanese War propaganda. He made speeches on the way through Yongchuan, Longchang, Neijiang and Ziyang. At 4 o'clock in the afternoon of 13, at the invitation of the Wartime International Rescue Team of Huaxiba Associated Universities in Chengtu, Feng Yuxiang in a military uniform came to Huaxiba to do a speech scheduled in the auditorium on second floor of the Whiting Memorial Administration Building. But there were more than two thousand students to listen to the speech, the auditorium can only accommodate two or three hundred people, so the speech had to be

moved to front ground of the Whiting Memorial Administration Building gate, and General Feng stood on a table to do speech.

冯玉祥对师生们说："今天诸位要知道我们能得到平平安安的在这里教书读书，自有是自广东起至大阴山内外蒙古一万多里路的战线上的三百万忠勇将士，用他们的血肉筑成一座新的长城，挡着日本鬼子过不来啊！才能在这里教书，才能在这里读书，我们应当怎样，应当如何地努力，这个书是怎样努力地读法。"

Feng Yuxiang said to the teachers and students, "Today everybody should know that you can safely teach students and study here, because on the battle line of over ten thousand miles from Guangdong to the inner Mengolia, more than three million brave soldiers build a new Great Wall with their flesh and blood, resisting the Japanese devils! Thus all of you can teach students here, can study here. What should you do? What you should do is studying hard."

抗战期间来华西坝采访的新闻记者里有一位特别的女性记者——她就是作家伊芙·居里，她的母亲居里夫人曾两次获得诺贝尔物理学奖。1942年春，艾芙·居里以战地记者的身份来到四川采访。在成都期间，她住在华西坝，不仅参

冯玉祥在怀德堂演讲后与五大学同学合影。左至右：熊德绍、文宝英、郭友文、张履建、唐波成、冯玉祥、张凌高、郭成圩、周曼如、张素芳。四川大学档案馆提供。

Feng Yuxiang took photo with students of the Associated Universities in Chengtu after his speech in the Whiting Memorial Administration Building. Left to right: Xiong Deshao, Wen Baoying, Guo Youwen, Zhang Lüjian, Tang Bocheng, Feng Yuxiang, Zhang Linggao, Guo Chengwei, Zhou Manru, Zhang Sufang. Photo source: Sichuan University Archives.

观了金陵女子文理学院在华西坝修建的简易女生宿舍，还在怀德堂里与坝上的女同学进行对话。艾芙·居里在她的文章中写道，在华西坝茶话会上有女同学对她说："我们本以为同盟国参战，几天就会把日本人赶出中国，可是事实令我们失望，不过我们有决心把抗战进行到底。"这位女生的发言赢得了众多学生的喝彩。艾芙·居里的报道足以见得中国人民抗日的决心，她把中国人民抗战到底的精神展现给了全世界。

During the Counter-Japanese War, there was a special female journalist who visited Huaxiba. She was the writer Eve Curie, her mother Madame Curie won the Nobel Prize in physics twice. In the spring of 1942, Eve Curie came to Sichuan as a war correspondent. During her stay in Chengdu, she lived in Huaxiba, visited the simple female dormitory built by Ginling College in Huaxiba, and talked with female students in the Whiting Memorial Administration Building. In her article, she wrote that a female student at the Huaxiba tea party told her, "We thought that the Allies entered the war would drive the Japanese out of China in a few days, but the fact disappointed us. But we are determined to fight to the end." The girl's speech won the applause of many students. The report of Eve Curie is enough to see the determination of the Chinese people to resist Japanese aggression and she showed the spirit of the Chinese people to the whole world.

1943年4月27日，时任英国驻华大使馆科学参赞、中英科学合作馆馆长李约瑟来到华西坝进行学术交流。从5月3日到5月24日，这位近代生物化学家、科学技术史专家在华西坝不同场地作了12场专题演讲，其中《战时与平时在英国之科学组织》《生命物质与灵魂》和《中国科学史与科学思想检讨》等6场讲演就是在怀德堂二楼礼堂举行的。据负责接待李约瑟的华西坝五大学"东西文化学社"干事郭祝崧回忆道："李博士声称，他将以华语讲课，可他是凭字典学的华语，因而我方学者们都劝他用英语讲，不过并不能促成他改变主意。"就这样李约瑟用他通过字典学的中文给五大学的师生们作演讲，"他在历次讲学时语言十分诙谐，不断引起全堂笑声，这样，许许多多深奥的内容，能够引起大家的兴趣。"郭祝崧说。

On April 27, 1943, Joseph Needham, the then scientific Counsellor of the British Embassy in China and the curator of Sino-British Scientific Cooperation Center, came to Huaxiba for academic exchange. From May 3 to May 24, this modern biochemist and an expert in the history of science and technology gave 12 lectures in different places of Huaxiba, among which six lectures, such as *Scientific Organization in Wartime and Peacetime in Britain*, *Living Matter and Soul*, and *Review of Chinese Scientific History and Scientific Thought*, etc., were held in the auditorium on the second floor of the Whiting Memorial Administration Building. Mr. Guo Zhusong, the executive director of the East-West Cultural Society of Huaxiba Associated Universities in Chengtu, who was in charge of receiving Needham,he recalled, "Dr. Li claimed that he would be lecturing in Chinese, but he learned Chinese through a dictionary, so our scholars advised him to speak in English, but this did not change his mind." It was in this way that Needham used the Chinese he had learned through the dictionary to give lectures to teachers and students of Associated Universities in Chengtu. "In his lectures, he spoke in very witty language, constantly causing laughter in the whole room, so that many esoteric content, can arouse everyone's interest." Mr. Guo Zhusong said.

华西坝老建筑的前世今生
The Stories of the Historic Architecture of Huaxiba

陆

西部图书馆与博物馆肇始

懋德堂

The Lamont Library and the
Museum Building, the Origin of West China
Library and Museum

怀德堂对面的这栋大楼叫懋德堂，1926年竣工，美国赖孟德夫妇为纪念其子捐建。该建筑落成后为大学的图书馆。1932年，华西协合大学博物馆成立，懋德堂的二楼作为古物博物馆。

The building opposite to the Whiting Memorial Administration Building is the Lamont Library and the Museum Building, completed in 1926, and donated by Mr. and Mrs. B.C. Lamont of U.S.A., in memory of their son. When completed, the building served as the university library. In 1932, the West China Union University Museum was established, and the second floor of the building served as an antiquities museum.

懋德堂在华西坝中轴线东侧，与事务所相对，其建筑形式与事务所相同，只是屋顶烟囱和老虎窗等西式建筑的元素减少了，整个造型的中式风格更加突出。

The Lamont Library and the Museum Building is on the east side of the central axis of Huaxiba, opposite the Administration Building on the west side of the central axis of Huaxiba. Its architectural form is the same as that of the Administration Building, but the symbols of western buildings such as roof chimneys and roof windows are reduced, and the Chinese style of the whole shape is more prominent.

懋德堂正面入口门廊屋顶为中国传统建筑的歇山式屋顶，柱基石、大圆柱以及屋顶的脊兽都很有特色。图书馆大阅览室的二楼是回廊，回廊上出现连续的西式半圆形拱券，立柱上雕刻有猫头鹰图案。

The front entrance porch of the Lamont Library and the Museum

1926年竣工的懋德堂，在照片右边是大楼的南侧，没有修建入口。图片来源：四川大学档案馆。

The Lamont Library and the Museum Building was completed in 1926. On the right is the south side of the building, with no entrance built. Photo source: Sichuan University Archives.

如今懋德堂为四川大学华西医学展览馆。戚亚男摄于2022年。
Now, the Lamont Library and the Museum Building is the West China Medical Exhibition Hall of Sichuan University. Photo by Qi Yanan in 2022.

Building is the saddle roof of traditional Chinese architecture. The pillar cornerstone, large columns and the ridge beasts on the roof are very distinctive. The second floor of the large reading room of the library is a indoor cloisters with continuous Western-style semi-circular arches, with owl patterns carved on the columns.

懋德堂屋脊的脊兽为二龙吐水，寓意避火；二龙中间为一古人用的书桌——几案，上有一本书，明确表明了此楼用作图书馆的功用。可见英国建筑师荣杜易是完全了解中国传统建筑中脊兽的作用——镇宅避邪，驱除妖孽，佑护平安。

The ridge beasts on the roof of the Lamont Library and the Museum Building are the two Loongs spitting water, which means to avoid fire. In the middle of the two Loongs is a desk used by ancient people, on which there is a book, which clearly indicates the function of the library. It can be seen that the British architect Rowntree fully understood the role of the ridge beast in traditional Chinese architecture, which could ward off evil, exorcise evil, bless and protect peace.

荣杜易手绘的懋德堂室内装饰猫头鹰的手稿。赵安祝提供。
The sketch of decorative owl for the reading hall of the Lamont Library and the Museum Building hand-painted by Mr. Rowntree. Andrew George provided.

20世纪50年代初的院系调整，华西协合大学的博物馆全部移交给四川大学。现在四川大学博物馆是目前中国高等院校中最重

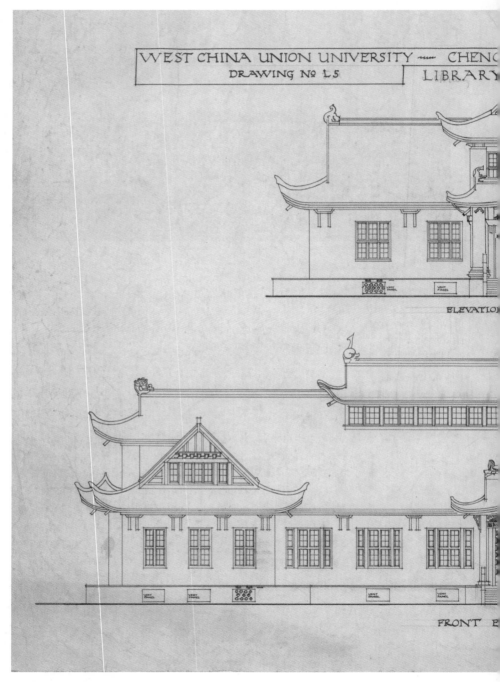

荣杜易设计的懋德堂侧翼和正面的立面图。赵安祝提供。
The wing and the front elevation of the Lamont Library and the Museum Building designed by Mr. Rowntree. Andrew George provided.

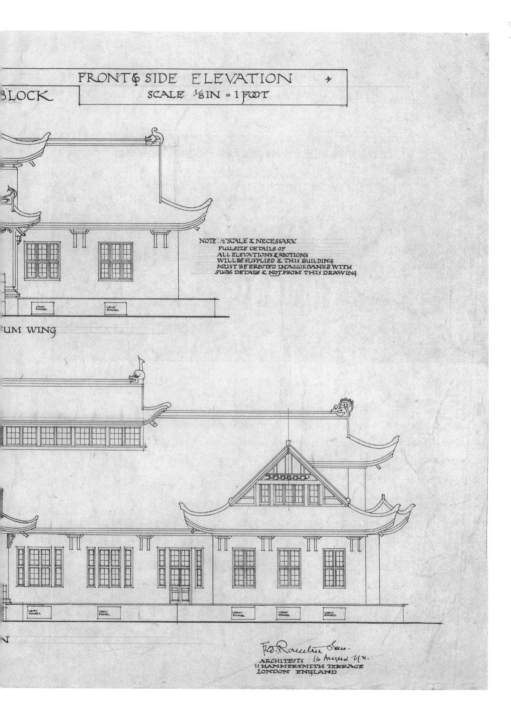

要的综合性博物馆之一。

After the department adjustment in the early 1950s, the museums of West China Union University were transferred to Sichuan University. Now Sichuan University Museum is one of the most important comprehensive museum among Chinese universities.

此后，懋德堂就是大学图书馆。1982年，在距懋德堂不到50米的地方新建了一座图书馆，新馆投入使用后，懋德堂就成了学校诸多行政办公所在地。

Since then, the Lamont Library and the Museum Building served only as the university library. In 1982, a new library was built less than 50 meters away from the old Library Building. Then the old one served many administrative offices of the university.

现如今，懋德堂为四川大学华西医学展览馆。2013年5月，懋德堂被国务院确定为全国重点文物保护单位。

Now, the Lamont Library and the Museum Building is the West China Medical Exhibition Hall of Sichuan University. In May, 2013, the Lamont Library and the Museum Building was confirmed as a National Key Cultural Relic Protection Unit by the State Council.

现为四川大学华西医学展览馆的懋德堂屋脊的脊兽依然是当年落成时的模样，一点都没有改变。戚亚男摄于2022年。

The Lamont Library and the Museum Building now is the West China Medical Exhibition Hall of Sichuan University. The ridge beasts on the roof of the building are still what they were when the building completed. Nothing has changed at all. Photo by Qi Yanan in 2022.

陆 · 西部图书馆与博物馆肇始——懋德堂

荣杜易设计的懋德堂。图片来源：1924年英国《建筑师》杂志。
The Lamont Library and the Museum Building designed by Mr. Rowntree.
Photo source: 1924 British magazine *The Builder*.

正在修建中的懋德堂。赵安祝提供。
The Lamont Library and the Museum Building under construction. Andrew George provided.

77

四川近代高校图书馆肇始
The First Modern University Library in Sichuan

作为一所大学，图书馆是其必不可少的机构。建校之初，华西协合大学没有固定的图书馆场所，藏书也不多。1916年11月，美国赖孟德夫妇向学校捐赠了15 000美元用于修建一栋永久性图书馆大楼。从购买土地、设计到修建，经过10年的时间，一栋建筑面积3 000多平方米的中西合璧的图书馆大楼落成了，该大楼是当时西部地区最完备的图书馆，一楼阅览大厅可容纳一两百人。

As a university, the library is an essential component. At the beginning of West China Union University's establishment, there was no fixed library place and only a few books. In November, 1916, Mr. and Mrs. Lamont donated $15,000 to the university for the construction of a permanent library building. From the purchase of land, design to construction, after 10 years of effort, a library building with a building area of more than 3,000 square meters erected, which was the most complete library in the western region at that time. The reading hall on the first floor could accommodate one or two hundred people.

落成后的懋德堂一楼大厅是图书馆阅览室，抬头可见梯形屋架。图片来源：四川大学档案馆。

The first floor of the Lamont Library and the Museum Building is the library reading hall. Looking up, there is trapezoidal roof frame. Photo source: Sichuan University Archives.

荣杜易手绘的懋德堂阅览大厅手稿,大厅有360平方米,层高14米。赵安祝提供。
The sketch of the reading hall of the Lamont Library and the Museum Building hand-painted by Rowntree. The hall is 360 square meters, 14 meters high. Andrew George provided.

抗战时期，图书馆为流亡到成都的学生求知发挥了很大的作用，图书馆的藏书也在此期间有了增加。当时南京金陵大学、金陵女子文理学院、济南齐鲁大学、北平燕京大学等学校先后内迁到成都与华西协合大学联合办学，各学校内迁时所带图书甚少，该图书馆对他们都开放，使得图书馆借阅书籍的人数激增，有资料显示，此时期每月平均有2.5万人次前去借阅图书。

During the Counter-Japanese War, the library played a great role for students in exile to Chengdu, and the library collection also increased during this period. When University of Nanking, Ginling College, Cheeloo University, Yenching University and other universities had successively moved to Chengdu and jointly run universities with West China Union University, they moved south with only a few books. The library's open to them made the number of borrowing books surge. Data showed that an average of 25,000 people borrowed books there every month at that time.

经过几十年的发展，到1949年，华西协合大学图书馆藏书量已有23万余册，其中外文图书有8万余册；有四川142个县的地方志共260余种，2 200多卷。订的中、外文杂志有500余种；收藏的成都市的日报，从1922年起至1949年完整无缺。

After decades of development, by 1949, the library of West China Union University had collected more than 230,000 books, including more than 80,000 books in foreign languages, more than 260 kinds and more than 2,200 volumes local chronicles of 142 counties in Sichuan. It had subscribed to more than 500 kinds of Chinese and foreign magazines and collected newspapers of Chengdu complete from 1922 to 1949.

古来华西第一馆
The First Museum in West China since Ancient Times

一所大学图书馆藏书的多少直接反映了该大学的办学实力，而一所大学博物馆藏品的多少则反映了该校的人文历史底蕴。华西协合大学博物馆是西南地区最早建立的博物馆，被誉为"古来华西第一馆"。早在1914年理学院美籍教授戴谦和等人就开始收集古物，至1931年已收集有6 000余件，藏品颇为可观，均陈列在懋德堂楼里。1932年，博物馆正式迁入图书馆二楼，馆长为美国人类学家葛维汉，从此博物馆的发展进入了新时期。其后中国考古学家郑德坤出任馆长，他制订的两个五年计划使博物馆更上一层楼，也使该馆成为成都的一处著名的文化圣地。有资料显示，1941年至1947年，博物馆馆藏文物已有3万件，先后来馆参观的人士不下数十万。

The number of books in a university library directly reflects the strength of the university, while the museum of a university reflects the cultural and historical deposits of the university. The Museum of West China Union University is the earliest museum established in southwest China, known as "The First Museum in West China since Ancient Times". As early as 1914, Dr. Daniel

懋德堂二楼为华西协合大学的古物博物馆，照片里显示的展柜里和柜顶上都是博物馆收藏的瓷器。图片来源：耶鲁大学。

On the second floor of the Lamont Library and the Museum Building is the Antiquarian Museum of West China Union University. The photo shows the collected porcelain in the showcase and on the top of the showcase. Photo source: Yale University.

Sheets Dye, an American professor of the College of Science, and others began to collect antiquities. By 1931, they had collected more than 6,000 pieces. The collection was considerable, and displayed in the Hart College (the College of Science). In 1932, the museum officially moved to the second floor of the library building. The curator was American anthropologist David Crockett Graham, and from then on, the museum entered a new period of development. Then curator Zheng Dekun, a Chinese archaeologist, took the museum to the higher achievement with two five-year plans, making it a famous culture place in Chengdu. Some data shows that from 1941 to 1947, the museum's collection of cultural relics had reached 30,000 pieces, and more than 100,000 people had visited the museum successively.

1934年，葛维汉馆长在得到了四川省教育厅的发掘执照后，组织了对四川广汉三星堆遗址的首次考古发掘，发掘到的几百件各种玉器、石器和陶器都收藏在华西协合大学博物馆内，由此揭开了以后半个多世纪的三星堆遗址考古序幕。

In 1934, after obtaining the excavation license of the Department of Education of Sichuan Province, Curator Graham organized the first archaeological excavation of the Sanxingdui Site in Guanghan, Sichuan Province, and hundreds of pieces of jade, stone ware and pottery excavated were also collected in the museum. Thus the next half a century of Sanxingdui Site archaeological prelude was opened.

1947年6月,四川邛崃西河洪水暴涨,河岸土层因洪水冲击垮塌,冲出来不少佛教石刻残件。不久,一些出土的佛经石刻拓片流传到了成都,引起了华西协合大学博物馆的注意,博物馆馆长葛维汉派馆员成恩元前去发掘。成恩元发掘、采集回来的石刻佛像雕刻十分精美,特别是那尊高1.98米的断臂观音佛像,发现时已断成4截,且散落在500米远的河床上。观音佛像经修复后重新站立起来,可是两只手臂已经不翼而飞。这尊断臂观音佛像被考古界誉为"中国维纳斯",是华西协合大学博物馆的"镇馆之宝"。

In June, 1947, the flood of Western River in Qionglai, Sichuan province skyrocketed. The soil layer on the river bank collapsed due to the impact of the flood, and many Buddhist stone carvings residue were washed out. Soon, some unearthed stone rubbings of Buddhist sutra were spread to Chengdu and attracted the attention of the museums. The curator of the university museum, Mr. Graham sent staff member Mr. Cheng Enyuan to explore. The stone Buddha statues excavated and collected by Cheng Enyuan are very exquisite, especially the 1.98 meter high broken arm Guanyin Buddha statue, which was found broken into

现为四川大学华西医学展览馆的懋德堂二楼,梯形屋架就呈现在眼前。戚亚男摄于2021年。
On the second floor of the Lamont Library and the Museum Building, now the West China Medical Exhibition Hall of Sichuan University, trapezoidal roof can be visible. Photo by Qi Yanan in 2021.

1934年，葛维汉（左边穿靴戴帽的外国人）在四川广汉三星堆遗址进行了首次考古发掘。图片来源：四川大学档案馆。
In 1934, Curator Graham(foreigner wearing boots and hat) conducted his first archaeological excavation at the Sanxingdui Site in Guanghan, Sichuan Province. Photo source: Sichuan University Archives.

four pieces and scattered on the riverbed one mile away. The Guanyin Buddha Statue stood up again after restoration, but both arms were missing. This statue is known as "Chinese Venas" by the archaeological circle, and is the "Unique Treasure of the Museum".

1952年，全国高校院系调整，华西协合大学的博物馆被调整至四川大学。

In 1952, the national universities and departments were reordered. The museum of West China Union University were transferred to Sichuan University.

百年来，四川大学博物馆积累了4万多件藏品，主要分为民族学、民俗学、石刻、书画、陶瓷、青铜器、古钱币、古印、刺绣、外国文物等门类。

For a hundred years, Sichuan University Museum has accumulated more than 40,000 pieces collection, which can be divided into ethnology, folklore, stone carving, painting and calligraphy, ceramics, bronze ware, ancient coins, ancient seals, embroidery, foreign relics and other main categories.

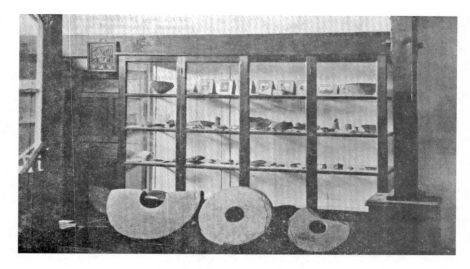

上图：这是四川大学博物馆展出的广汉三星堆在1934年首次考古发掘的各种玉器、石器和陶器等，最大的石璧直径51.4厘米。现在这些珍贵的文物收藏在四川大学博物馆内。图片来源：《华西边疆研究学会杂志》1934年，第6卷。

This is the museum's exhibition of various jade articles, stone tools and pottery which is first archaeological excavation of Sanxingdui, located in Guanghan, in 1934. The largest stone disk is 51.4cm in diameter. Now these precious cultural relics are collected in the Museum of Sichuan University.
Photo source: *Journal of the West China Border Research Society*, 1934, Vol.6.

下图：1947年6月，博物馆馆员成恩元在四川邛崃发掘的断臂观音佛像。易艾迪提供。

The broken arm Guanyin Buddha, excavated from Qionglai, Sichuan, in June, 1947, excavated by museum librarian Cheng Enyuan. Yi Eddie provided.

华西坝老建筑的前世今生
The Stories of the Historic Architecture of Huaxiba

自然科学的圣地

嘉德堂

The Atherton Building for Biology and Preventive Medicine, the Holy Land of Natural Science

上图：1924年嘉德堂竣工后在楼前举行庆典仪式，那时其右前边还没有钟楼。赵安祝提供。
When the Biology Building was completed in 1924, a celebration ceremony was held in front of the building. At that time, there was no clock tower in its right front. Andrew George provided.

下图：2012年威亚男拍摄的嘉德堂，照片右前边是1926年建成的钟楼。
The Biology Building. There is a clock tower built in 1926 in its right front. Photo by Qi Yanan in 2012.

位于钟楼旁的嘉德堂又名生物和预防医学楼，简称生物楼，1924年4月竣工，楼长59米、宽27米。该楼由美国夏威夷的嘉德尔顿医生夫妇为纪念其子亚历山大·阿瑟顿博士捐建。嘉德堂位于华西钟楼之东侧，原为生物学、生物化学、化学等教学场所，现为华西医学中心的解剖楼。

The Biology Building completed in April 1924, donated by Dr. Gardleton and his wife in Hawaii, U.S.A., in memory of their son, Dr. Alexander Atherton. The building is 59 meters long and 27 meters wide, located on the east side of the the clock tower. The Departments of Biology, Biochemistry, Chemistry, and the Natural History Museum used to be in this building. Now it is the anatomical building of West China Medical Center.

嘉德堂的门楼在一排斗拱下，是典型的中式木牌坊门楼，石栏杆后四根大圆柱的上端有四只木刻凤凰装饰物作为支撑木，檐口造型呈波浪形，屋脊的两端有造型夸张的动物。门楼

台阶和石栏杆由红砂石构成，栏杆的每一根望柱上都雕刻有一条龙，石栏板上刻有中国传统祥云图案。屋脊左右两头各有一对怪异的脊兽拥抱着一团祥云，寓意对生命的美好向往。

The gatehouse of the Biology Building is a typical Chinese wooden archway gate under a row of bucket arch. At the top end of the four large columns behind the stone railing there are four woodcut phoenix ornaments as supporting wood. The cornice shape is wavy, and there are animals with exaggerated shapes at both ends of the roof ridge. The gatehouse steps and stone balustrades are made of red sand stone. Each baluster is carved with a Loong, and the stone railing plates are carved with the traditional Chinese auspicious cloud pattern. On the left and right roof ridge each has a pair of weird ridge animals holding a auspicious cloud, meaning a good yearning for life.

20世纪50年代初的院系调整，嘉德堂里自然历史博物馆的标本全部被调整至四川大学。嘉德堂就成为医学院的人体解剖楼，被命名为第一教学楼，之后大家都称该楼为解剖楼。

In the early 1950s, the national universities and departments were reordered and all the specimens of the Natural History Museum in the Biology Building were transferred to Sichuan University. Since then the Biology Building only served the Human Anatomy Department of the Medical School, named the First Teaching Building, later it was called the Anatomy Building.

荣杜易手绘的嘉德堂。赵安祝提供。
Mr. Rowntree's hand-painted Biology Building. Andrew George Provided.

2001年11月，第一教学楼被成都市政府批准列为成都市首批文物建筑之一。

2002年12月，第一教学楼被四川省政府批准列为四川省第六批文物保护单位。

2013年5月，学校恢复命名，将第一教学楼改回嘉德堂；同年5月，该楼被国务院确定为全国重点文物保护单位。

In November, 2001, the Biology Building was confirmed as one of the first Cultural Relic Buildings in Chengdu by the Chengdu Municipal Government.

In December, 2002, the Biology Building was confirmed as the sixth batch of Cultural Relic Protection Units in Sichuan Province by the Sichuan Provincial Government.

In May, 2013, the Biology Building was confirmed as a National Key Cultural Relic Protection Unit by the State Council.

中国西南地区首家自然历史博物馆
The First Natural History Museum in the Southwest China

嘉德堂主要用于华西协合大学生物学系教学的场所，在该楼设立一所自然历史博物馆对于教学和研究是必不可少的，因而大楼投入使用时就成立了博物馆，该馆可谓西南地区首家自然历史博物馆。博物馆成立之初没有经费用于制作和购买标本，大部分标本为该系教职员工历年旅行所采集以及学生自告奋勇到野外去采集各种动植物制作而成，还有友人与相关学术单位提供。

The Biology Building is mainly used for the teaching of the university's Biology Department. It is essential to set up a natural history museum in the building for teaching and research, so the museum was established when the building was put into use, which can be said as the first natural history museum in southwest China. At the beginning of its establishment, the museum had no funds for the production and purchase of specimens. Most of the specimens were collected by the faculty and staff of the department during their travels and by students who volunteered to go out into the field to collect specimens of various animals and plants, as well as specimens provided by friendly people and related academic institutions.

经过前后近40年的积累，到20世纪40年代，华西协合大学自然历史博物馆的标本共有38 000多件，其中，哺乳类1 389件，鸟类1 554件，爬行类466件，两栖类1 740件，鱼类949件，昆虫19 378件，地质矿物类945件，古生物349件，植物10 902件，其他无脊椎动物500件。上列标本中两栖类的质量居当时国内之首，该馆所藏华西两栖类标本的数量当时在世界上也算丰富的。馆内标本按鸟类、哺乳类、冷血脊椎动物类、昆虫类、植物类、地质矿物类进行陈列。为了普及生物科学知识，该馆还向社会民众开放，有一年参观的人数曾达到1万人次。

After nearly 40 years of accumulation, by the 1940s, there were more than 38,000 specimens in the Natural History Museum, including 1,389 mammals, 1,554 birds, 466 reptiles,

嘉德堂二楼生物实验室里一批学生正在窗旁桌前观察显微镜下的标本，室内中间摆放着生物标本展柜，一位老师在实验室巡回走动便于指导学生。赵安祝提供。

A group of students in the biological laboratory on the second floor of the Biology Building are observing the specimens under the microscope at the table by the window. In the middle of the room are biological specimens showcases, and a teacher is walking around the laboratory to guide the students. Andrew George provided.

1,740 amphibians, 949 fishes, 19,378 insects, 945 geological minerals, 349 ancient creatures, 10,902 plants, and 500 other invertebrates. The listed specimens ranks the first in China with the quality of amphibian species, and the number of amphibian specimens collected in the museum in West China is also the most abundant. The specimens in the museum are displayed in categories of birds, mammals, cold-blooded vertebrates, insects, plants and geological minerals. The museum is also open to the public to popularize biological science knowledge, and one year the number of visitors reached 10,000.

1935年，蜀华中学的白仲山和程崇光两位同学参观了华西协合大学自然历史博物馆后，写了一篇杂记刊登在报纸上，他们是这样描写的：

In 1935, two students from Shuhua Middle School, Bai Zhongshan and Cheng Chongguang, visited the Natural History Museum of West China Union University and wrote an essay published in a newspaper. They described it as follows.

"出了图书馆，折向左转经过了一条小路和一所小桥到达了科学研究舍，里面陈列的是生物标本和化学仪器，关于生物标本确是使我们大饱眼福，增加了不少的见闻，恐怕全川的学校，再也没有这样充分的标本……"

"Out of the library, turn left through a path and a small bridge to the scientific research institute, where display biological specimens and chemical instruments, about biological specimens really feast our eyes, increased a lot of knowledge. I'm afraid the schools in Sichuan, no longer have such sufficient specimens…"

1952年全国院系调整，华西协合大学自然历史博物馆的标本全部被调整到四川大学了。

In 1952, the national universities and departments were reordered, and all the specimens of the Natural History Museum of West China Union University were transferred to Sichuan University.

茂汶苹果之父张明俊
Zhang Mingjun, the Father of Maowen Apple

2012年年底，农业部（现为中华人民共和国农业农村部）在北京召开的"2012年第三期中国农产品地理标志保护产品专家评审"会上，四川"茂汶苹果"通过了专家评审，成为国家地理标志保护产品。而这一切都源于张明俊——茂汶苹果之父。

At the end of 2012, the Ministry of Agriculture of the People's Republic of China (Now Ministry of Agriculture and Rural Affairs of PRC) held "2012 the Third Phase of China's Agricultural Products Geographical Indication Protection Product Expert Review" in Beijing. Sichuan's "Maowen Apple" passed the expert review and became the National Geographical Indication Protection Product. All this comes from Zhang Mingjun, the father of Maowen Apple.

1894年，张明俊出生于四川新都斑竹园，他小时候父亲就去世了，家境贫寒，靠母亲种地和做针线活维持生活。张明俊早年只读过私塾，后来在家务农。18岁那年，张明俊经舅父介绍，到成都华西协合高级中学当勤杂工。为了读中学，张明俊还到大学里去打工挣学费，如给生物系清洗制作动植物的标本。由于他做事认真、聪明好学，深得生物系教授白明道的喜爱。中学毕业后，张明俊被保送入华西协合大学理学院生物系读书。1924年，张明俊毕业后留校任教。

Zhang Mingjun was born in 1894 in Banzhuyuan, Xindu County, Sichuan Province. His father died when he was a child. His family was poor, relying on his mother in farming and sewing to maintain life. Zhang Mingjun only attended private schools in his early years, and later farming at home. At the age of 18, Zhang Mingjun was introduced by his uncle to Chengdu West China Union Senior High School as a handyman. In order to go to middle school, Zhang Mingjun also took part-time jobs at the university to earn tuition. He cleaned and made animal and plant specimens for the biology department. Because of his earnest work, intelligent and studious, he won the love of the biology professor P. M. Bayne. After graduating from middle school, Zhang Mingjun was recommended to the Department of Biology, School of Science, West China Union University. In 1924, Zhang Mingjun taught at the school after graduation.

20世纪30年代华西协合大学理学院生物系张明俊（左一）和同事、学生在嘉德堂楼前合影。图片来源：四川大学档案馆。
Zhang Mingjun (left 1), colleagues and students of the Department of Biology of the School of Science, West China Union University, took a photo in front of the Biology Building in the 1930s. Photo source: Sichuan University Archives.

张明俊深知四川地区虽然种植了不少果树，但品种不够丰富，品质也不是最好的，因而为了引进新品种和改良水果的品质，他创办了"明明果园"，引进并培育出无核蜜橘、蟠桃、脐橙、柠檬、水蜜桃和樱桃等多种水果。然而张明俊从加拿大引进的苹果优良品种"金帅"和"黄皇后"在果园里试验培育效果却不理想。四川茂汶的赵吉昌20世纪30年代就在张明俊的果园里学习果树种植技术，后来赵吉昌的邻居陈世五也结识了张明俊，三人一商量，决定另寻地方继续试验苹果树的栽培。1940年，他们来到赵吉昌和陈世五的家乡茂汶考察，结果发现当地的土壤、气候、环境等条件比成都地区更适宜种植苹果树。随后，张明俊在成都负责引进品种优良的苹果树苗，赵吉昌和陈世五负责苹果树苗在茂汶的栽培和管理。引进的苹果树苗在茂汶生长得非常好，三年后开花结果，其果实的色、形、味都俱佳。慢慢地苹果种植就在茂汶发展起来了，并且扩大到阿坝藏族羌族自治州各个县及凉山彝族自治州部分地区，形成了一个新的苹果品牌——茂汶苹果。

Zhang Mingjun knew that although many fruit trees were planted in Sichuan, the varieties are not rich enough and the quality is not the best. So in order to introduce new varieties and improve the quality of fruit, he founded "Mingming Orchard", introduced and cultivated a variety of fruits such as seedless orange, flat peach, navel orange, lemon, water peach and cherries. However, Zhang Mingjun introduced excellent apple varieties Jin Shuai and Empress Huang from Canada, in the orchard experiment cultivation, the effect was not ideal. Zhao Jichang, from Maowen County, Sichuan Province, learned fruit tree planting techniques in Zhang Mingjun's orchard in the 1930s. Later, Zhao Jichang's neighbor Chen Shiwu also met Zhang Mingjun, and the three decided to find another place to continue to test the cultivation of apple trees. In 1940, they visited Zhao Jichang and Chen Shiwu's hometown Maowen County, and found that the local soil, climate, environment and other conditions were more suitable for apple planting than in Chengdu. Subsequently, Zhang Mingjun was responsible for introducing some of the excellent varieties of apple saplings in Chengdu, then Zhao Jichang and Chen Shijun were responsible for the cultivation and management in Maowen. The introduced apple trees grew very well in Maowen, and three years later,blossomed and bore fruit,with a good color, shape and taste. Gradually, apple planting developed in Maowen, and expanded to each counties of Aba Prefecture and parts of Liangshan Prefecture, forming an apple brand — Maowen Apple.

植物病理学家何文俊
Plant Pathologist He Wenjun

何文俊1909年8月30日出生于重庆巴县，后来全家搬到重庆市，靠父母做小生意维持生活。何文俊小时候就读于教会在重庆曾家岩办的求精中学，由于成绩优异，被保送进入华西协合大学读书。当时因家庭经济困难，何文俊差一点就失去了上大学的机会。

He Wenjun was born in Baxian County, Chongqing, on August 30,1909. Later, his family came to Chongqing City to maintain a life by their parents doing a small business. When He Wenjun was a child, he attended the Qiu Jun Middle School ran by the church in Zeng Jiayan, Chongqing. Due to his excellent grades, he was recommended to attend West China Union University. At that time, because of the family financial difficulties, He Wenjun almost lost the opportunity to go to university.

何文俊读书时理学院生物系的师资力量很弱，只有一位外籍教授和前几届留校的毕业生，在何文俊毕业留校那年，生物系仅有的一位外籍教授也离开了，整个生物系全靠毕业留校的中国同事开展教学和研究工作。当时生物系除了为本系开设课程外，还要为医学院、牙学院和药学系等其他院系开课。

When He Wenjun studied in the university, the teachers in the Department of Biology were less. There were only one foreign professor and a few former graduates. When He Wenjun graduated, the only foreign professor in the Department of Biology also left. The whole Department of Biology relied on Chinese graduated students to carry out teaching and research work. At that time, the Department of Biology not only offered courses for the Biology Department, but also for other Departments,such as Medical School, Dental School and Pharmacy Department.

20世纪30年代,华西协合大学理学院生物系何文俊(二排右二)、张明俊(二排右一)等师生与毕业学生在嘉德堂楼前合影。图片来源:四川大学档案馆。
In the 1930s, He Wenjun (second row right 2) and Zhang Mingjun (second row right 1) and others of the Department of Biology of the School of Science, West China Union University, took a photo with graduates in front of the Biology Building. Photo source: Sichuan University Archives.

1935年,何文俊到加拿大多伦多大学攻读生物博士学位,然而一年后他却转赴美国爱荷华州州立大学,师从国际著名的植物病理学家梅尔哈斯教授,攻读植物病理学。何文俊深知植物病理学是一门专门研究植物病害发生和发展的规律以及防治的科学,掌握了这门科学对于我们这样一个农业大国的农作物产量的提高是有重要意义的。

In 1935, He Wenjun went to the University of Toronto to study for a doctorate in biology, but a year later he moved to the Iowa State University, studying plant pathology under professor I. E. Melhus, an internationally renowned plant pathologist. He Wenjun knew that plant pathology is a science that specializes in studying the occurrence and development of plant diseases as well as their prevention and treatment, and mastering this science will have significant for the improvement of crop yield in such a big agricultural country as ours.

1941年,何文俊学成回国后任华西协合大学生物系教授兼主任,他在校内开设并讲授了"农业概论""遗传学""藻、菌形态学"和"植物病理学"等课程。1946年,国立四川大学聘请何文俊为该校植物病虫害系教授兼主任,他为该校开设并讲授了"植物病理学""真菌学""植病研究方法"和"高级植病技术"课程。

In 1941, after returning to China, He Wenjun became professor and director of the Department of Biology of West China Union University. He taught Introduction to Agriculture, Genetics, Morphology of Algae and Fungus, Plant Pathology, etc. in the university. In 1946, National Sichuan University hired He Wenjun as professor and director of the Department of Plant Diseases and Insect Pests. He taught Plant Pathology, Plant Disease Research Methods and Advanced Plant Disease Technology in Sichuan University.

1950年，中央政府要在西南地区建一所农学院，西南军政委员会文教部委员会专门调何文俊到重庆去主持筹建西南农学院。1953年9月18日，经中央人民政府委员会第28次会议通过，任命何文俊担任西南农学院副院长并颁发了由国家主席毛泽东签名的任命通知书。

In 1950, the Central Government would build an agricultural college in the southwest region, and the Committee of the Culture and Education Department of the Southwest Military and Political Committee specially transferred He Wenjun to Chongqing to preside over the establishment of the Southwest Agricultural College. On September18,1953, approved by the 28th Meeting of the Central People's Government Committee, He Wenjun was appointed as Vice President of the Southwest Agricultural College and issued a notice of appointment signed by the State Chairman Mao Zedong.

两栖爬行动物学家刘承钊
Amphibian and Reptile Expert Liu Chengzhao

抗战时期，嘉德堂接纳了苏州东吴大学生物系刘承钊教授，使得他的两栖爬行动物研究工作不仅没有中断，而且在四川还得到了更大的发展。他在峨眉山发现新属种胡子蛙，并且成为我国两栖爬行动物学的主要奠基人之一，日后世界两栖爬行动物学会秘书长、美国生物学会秘书长、美国生物学家克莱格莱主编的《两栖爬行动物发展史的丰碑》一书，记载了自16世纪以来已故世界著名两栖爬行动物学家151人，刘承钊是中国唯一的入选者。

During the Counter-Japanese War, Professor Liu Chengzhao from the Department of Biology of Soochow University exiled to West China Union University and worked in the Biology Building. At here his research on amphibians and reptiles was not only continued, but also developed further. He discovered bearded frog, a new amphibians genus and species on Mount Emei and became one of the main founders of Amphibian and Reptile Science in China. The book *Monuments in the History of Amphibian and Reptile Development* edited by Kraigadlei, Secretary General of the World Society of Amphibians and Reptiles, Secretary General of the American Biological Society and American Biologist, recorded 151 deceased world famous amphibian and reptilian scientist since the 16th century, and Liu Chengzhao was the only Chinese selected.

2016年,在华西坝上的图书馆前的草坪上立了一尊两栖动物和爬行动物专家刘承钊教授铜像。咸亚男摄于2016年。

In 2016, a bronze statue of Professor Liu Chengzhao, amphibian and reptile expert was built on the lawn in front of the library in Huaxiba campus. Photo by Qi Yanan in 2016.

刘承钊1900年出生于山东泰安县一个耕读世家。1913年,刘承钊进入泰安县教会办的萃英小学读书,由于家境贫寒,他曾两次辍学回家务农。1924年,刘承钊考入北平燕京大学心理学系,但因对动物学产生浓厚兴趣,第二年他便转入生物系学习,毕业后留校任教并从事两栖爬行动物的研究工作。

Liu Chengzhao was born in 1900 in Tai'an County, Shandong Province, at a farming and reading family. In 1913, Liu Chengzhao entered Cuiying Primary School run by the Church of Tai'an County. Due to his poor family needed him to go back home to farming, He dropped out of school twice. In 1924, Liu Chengzhao was admitted to the Department of Psychology of Yenching University in Peking. However, due to his strong interest in zoology, he transferred to the Department of Biology the following year. After graduation, he stayed to teach and engage in amphibian and reptile research.

1932年,刘承钊获美国洛克菲勒基金奖学金赴美国康奈尔大学攻读博士学位。两年后,刘承钊获博士学位并回国,在苏州东吴大学任教,继续从事两栖爬行动物的研究。

In 1932, Liu Chengzhao was awarded the Rockefeller Foundation Scholarship and went to Cornell University to study for his doctorate. Two years later, Dr. Liu Chengzhao received his doctorate and returned to China to teach at Soochow University in Suzhou, continuing to engage in amphibian and reptile research.

1937年，全面抗日战争爆发，东吴大学被迫迁到浙江湖州。年底日军逼近，时任东吴大学生物系教授兼主任的刘承钊与同事征得校方同意后带领18名学生投靠成都华西协合大学生物系继续他们的学业。

In 1937, when the Counter-Japanese War broke out in an all-round, Soochow University was forced to move to Huzhou, Zhejiang Province. At the end of the year, when the Japanese army was approaching, Dr. Liu Chengzhao, then professor and director of the Department of Biology at Soochow University, and his colleagues got permission from the university to lead 18 students to the Department of Biology at West China Union University in Chengdu to continue their studies.

1938年暑假，刘承钊带领学生到峨眉山进行为期两个月的野外动物采集，他第一次听到并采集到著名的"仙姑弹琴蛙"。一天，一位学生捕获了一只奇特的蛙，刘承钊经过多年的研究确定它是蛙类中的一个新属、新种，并把该蛙命名为"峨眉髭蟾"，俗称"胡子蛙"，这是第一次被中国科学家记录下来的一个新属种。

In the summer vacation of 1938, Dr. Liu Chengzhao led his students to Mount Emei for a two-month field animal collection. He knew for the first time and collected the famous "Fairy Frog". One day, a student caught a strange frog. After years of research, Dr. Liu Chengzhao identified it as a new genus and species of frog, and named the frog "Emei Moustache Toad" (*Vibrissaphora boringii* Liu), commonly known as "bearded frog". This is the first time that a new species has been recorded by Chinese scientists.

1946年至1947年，美国芝加哥自然历史博物馆两栖爬行动物部邀请刘承钊到美国进行学术交流。此期间刘承钊在芝加哥自然历史博物馆撰写了一部学术专著《华西两栖类》，该书1950年出版后，在国际两栖爬行学界引起极大反响，至今仍被视为研究中国两栖动物的经典著作。

From 1946 to 1947, the Amphibian and Reptile Department of the Natural History Museum in Chicago invited Dr. Liu Chengzhao to the United States for academic exchange. During this period, Liu Chengzhao wrote an academic monograph on *Amphibians of Western China* at the Chicago Museum of Natural History. The book was published in 1950 and aroused a great response from the international amphibian and reptile community. It is still regarded as a classic study of amphibians of China.

1950年，刘承钊应聘到北京燕京大学担任该校生物系主任。第二年，西南军政委员会文教部又把刘承钊从北京请回成都担任政府接管后的华西大学第一任校长。1952年全国院系调整后，华西大学改名为四川医学院，刘承钊改任院长。

In 1950, Dr. Liu Chengzhao was recruited to Beijing Yenching University as the Director of the Department of Biology. The following year, the Culture and Education Department of the Southwest Military and Political Committee invited Dr. Liu Chengzhao back to Chengdu from Beijing to serve as the first President of West China University after the government took over. In 1952, the national universities and departments were reordered and the university was renamed Sichuan Medical College and Dr. Liu Chengzhao became the dean.

华西坝老建筑的前世今生
The Stories of the Historic Architecture of Huaxiba

捌

抗战时期的建筑文物

懿德堂

Stubbs' Memorial Chemistry Building, the Building of the Counter-Japanese War Age

懿德堂原为化学楼，又名苏道璞纪念堂，由华西协合大学加拿大籍建筑工程师苏继贤负责建造。该楼1939年春开工，1941年秋竣工，位于生物楼对面，其外观与生物楼相似。为了纪念1930年在华西坝遇害的华西协合大学副校长、化学教授苏道璞，该楼被命名为苏道璞纪念堂。20世纪50年代，化学楼被命名为第二教学楼。

Stubbs' Memorial Chemistry Building was originally a Chemistry Building, also is known as Stubbs' Memorial Hall. The Canadian construction engineer of West China Union University, Walter Small was responsible for the construction. The building began in the spring of 1939 and was completed in the autumn of 1941, located opposite the Biological Building, with a similar appearance to it. It was also named Stubbs' Memorial Hall in honor of Clifford M. Stubbs, Vice President and chemistry professor of West China Union University, who

化学楼的外观与对面的生物楼基本相同，只是门楼台阶没有石栏杆。化学楼前还有一中式八角小亭附属建筑物，专门用于储放易燃易爆的实验物品，以确保化学楼的安全。图片来源：耶鲁大学。

The Chemistry Building located opposite the Biological Building, with a similar appearance to it, but the gatehouse steps have no stone railings. There is a Chinese octagonal pavilion in front of the building which was specially used for storing inflammable and explosive experimental items to ensure the safety of the chemical building. Photo source: Yale University.

was killed in Huaxiba in 1930. In the 1950s, the Chemistry Building was named the Second Teaching Building.

2001年11月，第二教学楼被成都市政府批准列为成都市首批文物建筑之一。

2002年12月，第二教学楼被四川省政府批准列为四川省第六批文物保护单位。

2013年5月，第二教学楼被国务院确定为全国重点文物保护单位。

In November, 2001, the Second Teaching Building was confirmed as one of the First Cultural Relic Buildings in Chengdu by the Chengdu Municipal Government.

In December, 2002, the Second Teaching Building was confirmed as the sixth batch of Cultural Relic Protection Units in Sichuan Province by the Sichuan Provincial Government.

In May, 2013, the Second Teaching Building was confirmed as a National Key Cultural Relic Protection Unit by the State Council.

现在楼前小亭已废用，它被七里香藤蔓所覆盖，外观仅可看到亭顶和一檐角。戚亚男摄于2012年。
Now the pavilion has been abandoned. It is covered by the Chinese daphne odera vine. The appearance can only be seen the roof and an eaves corner. Photo by Qi Yanan in 2012.

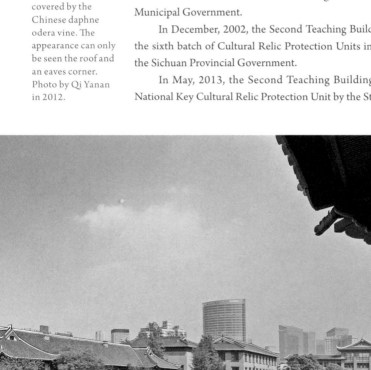

为延续中国高等教育事业的发展而建造的教学楼
A Teaching Building Built for Continuing the Higher Education

  1937年全面抗日战争爆发，战区大学纷纷内迁，华西协合大学也接纳了多所大学，为延续中国高等教育事业的发展，这些大学在华西坝联合办学。尽管华西协合大学倾尽全力为各校提供办学条件，各大学甚至采用联合办学的方法来解决教学资源的不足，且各大学的学生可跨校选课读书，各学校承认其所读学分，但就这样教学场地还是不足，特别是化学教学与实验室。1939年，由华西协合大学和抗战时内迁来华西坝的金陵大学、齐鲁大学和金陵女子文理学院出资并得到美国"中国基督教大学联合董事会"的资助而兴建了一座化学楼，当时商定，该楼建成后由四校的化学系及金陵大学的化工系合用，战后归华西协合大学所有。

  In 1937, when the Counter-Japanese War broke out in an all-round way, many universities in war zones moved to western China cities. West China Union University also accepted many universities. These universities jointly ran schools here, continuing the higher education of China. West China Union University had tried its best to provide educational conditions for each university, and they even took the method of joint school running to solve the shortage of teaching resources. Students from various universities could choose courses across schools, and the school admitted their credit they have read, but the teaching space was still insufficient, especially chemistry teaching

左图：化学楼的地下室也是实验室，同学们正在老师的指导下做实验。图片来源：耶鲁大学。
The basement of the Chemistry Building is also laboratory. The students are doing the experiments under the guidance of the teacher. Photo source: Yale University.

右图：化学楼里设有提供给师生们查阅专业图书、资料的图书室。图书室以苏道璞的名字命名，图书室椅子背上都刻有他的名字以示纪念。图片来源：耶鲁大学。
In the Chemistry Building, there is a library for teachers and students to consult professional books and materials. The library is named after Dr. C. M. Stubbs, and his name was engraved on the back of the chair in the library in memory of him. Photo source: Yale University.

and laboratory. So in 1939, a chemical building was constructed. It was jointly funded by University of Nanking, Cheeloo University, Ginling College and West China Union University, and also get funding from US Christian University Joint Board in China. At that time, it was agreed that the building after completion, was jointly used by the Chemical Department of four universities and the department of Chemical Industry of Ginling College, and owned by the West China Union University after the war.

1941年9月1日下午，在化学楼举行了大楼的落成典礼，到会的有四川省政府主席张群，当时正值中国化学会在该校举行第九届年会，四校联合邀请全体年会代表参加此次典礼，此外还有各界来宾及四校校级领导等数百人。

On the afternoon of September 1, 1941, the completion ceremony of the building was held in the Chemistry Building, attended by Zhang Qun, Chairman of Sichuan Province. At that time, the ninth Annual Meeting of the Chinese Chemical Society was held at the university. The authorities of the four universities invited all the representatives of the annual meeting to attend the ceremony, as well as guests from all walks of life and the leaders of the four universities.

在典礼上，首先由金陵大学陈裕光校长介绍了苏道璞纪念堂修建的原因及其意义，接着华西协合大学化学系徐维理主任给大家讲了苏道璞的个人经历。

At the ceremony, President Chen Yuguang of University of Nanking first introduced the origin and significance of the construction of the Chemistry Building. Then William Gawan Sewell, Director of Chemistry Department of West China Union University, told everybody Stubbs' personal experience.

苏道璞1888年11月出生于英格兰，21岁获新西兰大学学士学位，后获得英国利物浦大学理科硕士和博士学位。1913年，苏道璞被英国公谊会派到成都华西协合大学担任化学教授。他创办了化学系，并亲自编写讲课教材与实验教程，从英国引进了先进的化学教学设备和器材，曾担任理学院院长和大学的副校长一职。

Stubbs was born in November, 1888 in England. At the age of 21, he received a bachelor degree from the University of New Zealand, a master and a doctor degree of Science from the University of Liverpool in England. In 1913, Stubbs was sent by the British Quaker to Chengdu West China Union University as a professor of chemistry. He founded the Department of Chemistry, wrote teaching materials and experimental courses by himself, and introduced advanced chemistry teaching equipment from Britain. He served as dean of the College of Science and vice president of the university.

苏道璞是一位具有博爱之心的老师，他从来不坐轿子，他曾说："坐轿子很不人道。我不愿把中国人当作牛、马一样使唤。除非有病走不动，否则我永远不会坐轿子。"为此苏道璞专门从英国买来了一辆自行车，但谁也想不到这辆自行车会给苏道璞带来灭顶之灾。

Dr. Stubbs, a teacher with a philanthropic heart, never rode in a sedan chair. He once said, "It's not humane to ride in a sedan chair. I don't want to treat the Chinese like cattle or horses. I will never ride in a sedan chair unless I am too sick to walk." For this reason, Dr. Stubbs bought a bicycle from Britain, but no one could have thought that this bicycle brought disaster to Stubbs.

1930年5月30日晚，苏道璞在校内骑自行车经过合德堂前时，遭遇抢劫而被杀伤，因抢救无效，于6月1日死亡。苏道璞在去世前一天还告诉他的夫人请通过学校转告中国政府，不要因他受重伤而引起中英两国关系恶化。事后，四川省军警团联合办事处抓获了3名凶手，并依法枪决3名劫匪。

On the evening of May 30, 1930, Dr. Stubbs rode the bicycle past the Hart College in campus. He was robbed and killed, the rescue is invalid, and he died on June 1. But the day before his death, Stubbs told his wife to tell the Chinese government through the university that relations between China and Britain should not deteriorate caused by his serious injury. Later, the Joint Office of the Sichuan Provincial Military and Police Regiment seized three murderers and executed three robbers according to law.

徐维理说："此化学馆之建筑，乃象征（苏道璞）博士之伟大牺牲精神，凡在此工作及学习之一切男女，务体此旨，以博士之仁爱为怀，社会人类之前途，左图：2013年在英国出版发行了《几乎是一个中国人——苏道璞的一生》一书。
In 2013, the book *Nearly A Chinese: A Life of Clifford Stubbs* was published in the UK.

右图：当年化学系毕业同学赠送的刻有"所过者化"四字匾额后来丢失了。1985年75周年校庆时，化学系校友særreg重刻此匾，"所过者化"四字寓意要不断地弘扬化学事业。匾额现悬挂于化学楼门廊后，拾级而上回头可见。戚亚男摄于2022年。
The plaque that the Chemistry Department graduates gave as a gift was lost later. On the 75th anniversary of the university in 1985, alumni of the Chemistry Department specially re-engraved this plaque.The big four Chinese characters on the plaque means that there will always be people who develop chemical science . The plaque is now hung behind the porch of the Chemistry Building, which can be seen after you enter the door and look back.
Photo by Qi Yanan in 2022.

实利赖之。"

W. G. Sewell said, "This Chemistry Building is a symbol of the great sacrifice spirit of Dr. Stubbs. All men and women who work and study here should understand this great spirit and keep the benevolence of Dr. Stubbs as the future of social mankind depends on it."

四川省政府主席张群及化学会曾昭抡先后致辞，华西协合大学化学系毕业代表熊正德代表同学献上祝词和赠送匾额，祝词是："抗战建国，理化时需，宏模讲求，复兴之基"。匾额上书"所过者化"。

Provincial Chairman Zhang Qun and Zeng Zhaolun from Chemical Society successively delivered speeches, and graduation representative Xiong Zhengde of Chemistry Department of West China Union University presented a best wishes speech and gave a plaque gift on behalf of the students. He said, "The Counter-Japanese War and building the country needed science, studying and developing physics

and chemistry is the foundation of revival." There are four characters on the plaque, which mean there's always someone developing the science of chemistry.

建筑工程师苏继贤在修建化学楼时，为了保障大楼的安全，在该楼的前面修建了一幢中式八角小亭作为附属建筑物，专门用于储放易燃易爆的实验物品。现在小亭已废用，它被七里香藤蔓所覆盖，外观仅可看到亭顶和一檐角。

In order to ensure the safety of the Chemistry Building, construction engineer Walter Small also built a Chinese octagonal pavilion in front of the building as an attached building, which was specially used for storing inflammable and explosive experimental items. Now the pavilion has been abandoned, it is covered by the Chinese daphne odera vine, the appearance can only be seen the roof and an eaves corner.

吕锦瑷研制出了中国的第一张X光片
Lü Jinai Developed China's First X-ray Film

1941年，四所大学联合修建的化学楼投入了使用，华西协合大学化学系聘请金陵女子文理学院毕业生吕锦瑷担任讲师。吕锦瑷1912年出生于山西交城，她在北平读贝满女中期间，有一天，摄影家魏守忠来贝满女中给同学们照相，吕锦瑷对照相充满了兴趣，当了解到我们国家还不能生产胶卷时，她就立下志向要造出中国人自己的胶卷。为了实现这一梦想，吕锦瑷高中毕业后，考入了南京金陵女子文理学院化学系并于1936年毕业。

In 1941, the Chemistry Building jointly built by four universities was put into use. The Chemistry Department of West China Union University hired Ms. Lü Jinai, a graduate of Ginling College, as a lecturer. Ms. Lü Jinai was born in Jiaocheng, Shanxi Province in 1912. When she was studying at Bridgman Girls' School, in Beiping, one day, photographer Wei Shouzhong came to Bridgman Girls' School to take photos of the students. Ms. Lü Jinai was full of interest in photography. When she learned that our country could not produce photosensitive film, she made up her mind to make Chinese people's own photosensitive film. For this dream, after graduating from high school, Ms. Lü Jinai was admitted to the Chemistry Department of Ginling College and graduated in 1936.

1937年全面抗日战争爆发，吕锦瑷随丈夫孙明经所在的金陵大学教育电影部南迁到四川重庆（现重庆市，为直辖市）。在重庆期间，吕锦瑷除了应聘在中学教书外，她把精力都用在了研制感光材料上。1941年，吕锦瑷应聘到华西协合大学化学系任教，在新建的化学楼里有完备的实验室和各种化学试剂，很利于吕锦瑷研制感光材料。时值抗战困难时期，当时华西协合大学医院里的X光片供应困难，院方得知吕锦瑷在研制感光材料，而且也已取得了初步的成果后，就与她合华西协合大学化学系吕锦瑷讲师在研究室里做课题实验。她正在用中国传统计量工具杆秤称实验材料,在她后面的实验桌上有一台天平秤。图片来源:耶鲁大学。
Ms. Lü Jinai, a lecturer of the Department of Chemistry of West China Union University, is doing experiments in the research laboratory. She is using a traditional Chinese steelyard to weigh the experimental materials, and behind her there is a balance scale on the experimental table. Photo source: Yale University.

作。后来,吕锦瑷成功研制了中国的第一张X光片。用她研制的中国的x光胶片所拍摄的第一张诊断骨折的X光照片,至今仍被保存在美国耶鲁大学图书馆中。1942年,吕锦瑷为华西坝五大学开设了"摄影化学"课程,这是在我国高等院校中首次开设的课程。

In 1937, When the Counter-Japanese War broke out in an all-round way, Ms. Lü Jinai moved to Chongqing, Sichuan Province(Now it is Chongqing, a municipality directly under the Central Government.) with the Educational Film Department of Ginling University where her husband Mr. Sun Mingjing worked. In addition to teach in the middle school during the period of Chongqing, she devoted all her energy to develop photosensitive materials. In 1941, Ms. Lü Jinai was employed to teach in the Chemistry Department of West China Union University. There were complete laboratories and various chemical reagents in the newly built Chemistry Building, which was very beneficial for Ms. Lü Jinai to develop photographic materials. It was a difficult time of the Counter-Japanese War, and the supply of X-ray film in the West China Union University Hospital was difficulty. The hospital learned that Ms. Lü Jinai was developing photosensitive materials and also has achieved preliminary results and cooperated with her. Ms. Lü Jinai successfully developed the domestic X-ray film. The first X-ray photo of a diagnosed fracture taken with her successfully developed Chinese X-ray film is still preserved in the Library of Yale University. The following year, Ms. Lü Jinai opened a Photographic Chemistry Course for Huaxiba Associated Universities in Chengtu, which was the first time offered in colleges and universities in China.

中国现代皮革工业之父张铨
Zhang Quan, the Father of China's Modern Leather Industry

1939年11月，华西协合大学校长张凌高与国立中央技艺专科学校校长刘贻燕联合聘请在美国学习制革学的张铨担任两校化学系和皮革专科的教授，月薪300银圆，这在当时传为佳话。1940年，张铨获得博士学位后，回到成都任教。

In November, 1939, President Zhang Linggao of West China Union University, and President Liu Yiyan of the National Central Technical College jointly hired Zhang Quan, who studied tanning in the United States, as a professor of chemistry and tanning in the two schools with a monthly salary of 300 silver dollars, which was admired by the mass at that time. In 1940, After received his doctorate, Zhang Quan returned to Chengdu for employment.

张铨1899年10月11日出生于浙江省仙居农家。张铨5岁在家乡读私塾，8岁时父亲送他到县里的洋学堂读小学。1913年张铨小学毕业，被教会保送到杭州之江大学附中就读。1921年，张铨考入了北平燕京大学制革系学习，1925年毕业留校任教，从事制革教学工作。

Zhang Quan was born on October 11,1899 at a farmer family of Xianju county, Zhejiang Province. Zhang Quan studied in a private school in his hometown at the age of 5. When he was 8, his father sent him to the county missionary school for primary school. In 1913, he graduated from primary school and was recommended to attached middle school of Hangchow University. In 1921, Zhang Quan was admitted to the tanning Department of Yenching University in Peking. In 1925, he graduated and stayed in the university engaged in tanning teaching.

1937年，张铨到美国俄亥俄州辛辛那提大学制革研究系攻读博士学位。1940年5月，张铨的

1999年我国皮革科技先驱、制革化学家、教育家张铨教授铜像在四川大学望江校区落成。咸亚男摄于2004年。
In 1999, a bronze statue of Professor Zhang Quan, a leather technology pioneer, leather chemist and educator in China, was built in Wangjiang Campus of Sichuan University.
Photo by Qi Yanan in 2004.

博士论文在美国皮革化学家学会第37届年会上宣读，引起了大家的关注。辛辛那提大学想让张铨留在美国，他们承诺给予其丰厚的待遇和优越的工作条件，但张铨还是婉言谢绝。1940年11月，张铨抵达成都。在华西协合大学化学系，张铨开设了"制革工程"和"蛋白质化学"两门课程。

In 1937, Zhang Quan went to the University of Cincinnati, Ohio, US, to study for doctorate in leather research. In May 1940, Zhang Quan's doctoral thesis was read at the 37th Annual Meeting of the American Society of Leather Chemists, which attracted great attention. The University of Cincinnati wanted Dr. Zhang Quan to stay in the United States. They promised him generous salary and excellent working conditions, but Zhang Quan politely declined. In November, 1940, Dr, Zhang Quan arrived in Chengdu. In the Chemistry Department of West China Union University, Dr. Zhang Quan opened two courses: Leather Engineering and Protein Chemistry.

抗战以来，我国的制革厂遭到日军巨大的破坏，皮革既是军需物资又是民用物品，在西南后方恢复制革工业尤为重要。要搞皮革生产，就要培养专业技术人才，张铨承担起了这一重任。皮革是一门实践性很强的学科，为此华西协合大学与四川省政府在华西坝联合建立了用于学生实验的制革工厂，由张铨负责。随后成立了四川省立成都制革科高级职业学校，首任校长徐士弘上任不久即去重庆大学任教，随后四川省教育厅厅长郭有守亲自出面邀请张铨担任制革校校长。

化学楼的门楼因年久失修，20世纪50年代被拆除，几十年都没有恢复原样，让人们以为大楼的入口原本就是这样设计的。咸亚男摄于2002年。
The gatehouse of the Chemistry Building was in disrepair and was demolished in the 1950s and was not recovered for decade. People would mistakenly think the appearance of entrance to the building on the picture was originally designed. Photo by Qi Yanan in 2002.

Since the Counter-Japanese War, China's tannery industry had been greatly damaged by the Japanese army. Leather was both military and civilian goods. It was particularly important to restore the leather industry in the southwest rear. To engage in leather production, it was necessary to cultivate professional and technical personnel, and Dr. Zhang Quan undertook this important task. Leather was a highly practical major, for this reason, West China Union University and the Sichuan Provincial Government jointly established a leather factory for student experiments in Huaxiba, which was responsible by Dr. Zhang Quan. Subsequently, Sichuan Provincial Chengdu Leather Technology Senior Vocational School was established. The first President Xu Shihong, took office and soon went to Chongqing University to teach. The Provincial Education Director Guo Youshou personally invited Dr. Zhang Quan to serve as the leather school president.

1952年全国高校院系调整后,张铨在四川建立了新中国高校第一个制革学专业,也就是现在的四川大学皮革学科。

In 1952, after the adjustment of the university departments nationally, Dr. Zhang Quan was responsible for the establishment of the first tanner major of New China in Sichuan, which is now the Leather Discipline of Sichuan University.

修复好的化学楼门廊与原来的不完全一样,增建了台阶上的石栏杆。戚亚男摄于2012年。
The restored Chemistry Building porch is not the same as exactly the original. Stone railings were added on the steps. Photo by Qi Yanan in 2012.

华西坝老建筑的前世今生
The Stories of the Historic Architecture of Huaxiba

玖

因市政建设而移建的学舍

万德堂

The Vandeman Memorial Hall, Once Relocated Due to Municipal Construction

万德堂又名万德门，1920年竣工，由美国万德夫妇所捐建，原为英语系教学楼，同时还提供50名学生住宿的宿舍。万德堂集教室和学生宿舍于一体，称之为学舍。万德堂平面为"T"形，屋顶采用了多种组合形式，该楼最大的特点是屋顶上建有一个两层的中式攒尖顶圆亭，入口为一中式牌坊与三个西式拱券相结合，门前有两对抱鼓石，其上各有一只麒麟。

The Vandeman Memorial Hall was completed in 1920. The building was donated by the Vandeman Couple in the United States. It was originally an English teaching building, as well as a dormitory for 50 students. The building integrates classroom and student dormitories and is called the Learning House. The Vandeman Memorial Hall is T-shaped in plane, and the roof adopts various combination forms. The biggest feature of the building is that there is a two-story Chinese spire circular pavilion on the roof. The entrance of the building is a combination of a Chinese memorial archway and three Western style arches. There are two pairs of drum stones in front of the building, each with a Kylin on it.

万德堂屋顶的小亭是很别致的，门楼的造型是中式的牌坊形式。赵安祝提供。
The small pavilion on the roof of the Vandeman Memorial Hall is very unique, and the shape of the gatehouse is in the form of Chinese archway. Andrew George provided.

华西协合大学的学舍有别于其他学校，大学"仿英国牛津大学体制，实行'学舍'制。即按参加大学组织的差会划分区域，每个区域各自构成一个'学舍'。每个差会负责兴建一幢或数幢教学楼和学生宿舍，并由各自管理。""学生入学除在大学注册外，还得在所住的学舍注册，并接受该差会所派的舍长和舍监的管理。实行这种制度的宿舍，可起学生公寓的作

1960年因成都市修建人民南路，万德堂移建于钟楼旁，屋顶上中式攒尖顶圆亭没有了，添加了飞马和龙等脊兽。戚亚男摄于2012年。

In 1960, due to the construction of South Renmin Road in Chengdu city, the Vandeman Memorial Hall was moved next to the clock tower. On the roof, the small circular pavilion with Chinese style spires was missing, and ridge animals such as pegasus and loong were added. Photo by Qi Yanan in 2012.

用。它对学生的生活管理有章可循，较为严格，减轻了大学的行政负担。而大学则能集中精力主要管理专业科系、教育计划实施、教学科研的后勤、学校的规划建设等。"

The student dormitory of West China Union University is different from that of other universities. The university "imitated the British Oxford University system and implements 'Learning House' System. The university divided landed to each missionary society which was membership in the university to set up a college. Then each of the missionary society built one or more teaching building and student dormitories on the land they get and managed the college by themselves." "In addition to register at the university, students are required to register at the house in which they live and are under the supervision of head of dormitory and housemaster assigned by the missionary society. The dormitory that carries out this kind of system can play the role of student apartment. It has rules and regulations for the management of students' life, which is relatively strict and reduces the administrative burden of university. The university, on the other hand, can concentrate on the management of major departments, implementation of education plans, logistics of teaching and research, and school planning and construction."

1960年，因成都市修建人民南路，该楼迁建于钟楼旁，屋顶上中式攒尖顶圆亭没有了，添加了飞马和龙等脊兽。20世纪50年代万德堂被命名为第六教学楼，现为四川大学华西药学院的教学楼。

In 1960, due to the construction of South Renmin Road in Chengdu City, the Vandeman Memorial Hall was moved to the side of the clock tower. On the roof, there was no longer the two-story Chinese spire circular pavilion, and ridge beasts such as pegasus and loong were added. In the 1950s, the Vandeman Memorial Hall was named as the Sixth Teaching Building, which is the teaching building of West China School of Pharmacy of Sichuan University.

2001年11月，第六教学楼被成都市政府批准列为成都市首批文物建筑之一。

2002年12月，第六教学楼被四川省政府批准列为四川省第六批文物保护单位。

2013年5月，学校恢复命名第六教学楼为万德堂；同年5月，该楼被国务院确定为全国重点文物保护单位。

In November, 2001, the Sixth Teaching Building was confirmed as one of the First Cultural Relic Buildings in Chengdu by the Chengdu Municipal Government.

In December, 2002, the Sixth Teaching Building was confirmed as the sixth batch of Cultural Relic Protection Units in Sichuan Province by the Sichuan Provincial Government.

In May, 2013, the school renamed the Sixth Teaching Building as Vandeman Memorial Hall; May of the same year, the building was confirmed as a National Key Cultural Relic Protection Unit by the State Council.

成都中轴线上的一座老建筑
An Old Building on the Central Axis of Chengdu

20世纪50年代初，成都进行市政建设，确定了以明代蜀王府为中心，修建一条南北贯通的道路，总称人民路，这也是成都的中轴线。万德堂正好位于中轴线上，修路肯定要被拆。成都文化名人流沙河这样写道："想起二十世纪五十年代之初，小说家李劼人，时任成都市副市长，分管城市建设，一日到布后街二号院（四川省文联机关所在地）偶然谈起本市街道建设规划，兴致高昂，洪声亮嗓，说要从贡院街（皇城坝）向南拆。三桥北街、三桥南街、红照壁、纯化街，都要一扫而光。再向南，拆城墙、越锦江，一直向南拆下去。要拆出一条几十米宽的大道来，作为新成都向南面发展的中轴线。李劼人激动地说：'到那时候，登上皇城，向南一望，好开阔的十里长街，那才是雄伟！'……你看，一条直线向南拆去，真有切蛋糕之豪爽。谁想过这一刀切下去，南城外的华西坝就要被永远地腰斩了……不数年间，拆街规划变成现实，腰斩果然执行。于是有了天府广场，以及人民南路。所幸的是蛋糕虽已切成两块，毕竟还是蛋糕。偶入坝区一游，仿佛风景依旧。"

In the early 1950s, Chengdu carried out municipal construction. It was determined that the former Ming Dynasty Shu Palace also known as the Imperial City as the center of Chengdu, to built a road running between the north and the south, known as Renmin Road, which is also the central axis of Chengdu. The Vandeman Memorial Hall was exactly located on the central axis of

左图：万德堂入口为一中式牌坊与三个西式拱券相结合，屋顶上建有一个两层的中式攒尖顶圆亭，特别引人注目。图片来源：耶鲁大学。

The entrance of the Vandeman Memorial Hall is a combination of a Chinese memorial archway and three Western style arches. On the roof there is a two-story Chinese spire circular pavilion, which is particularly striking. Photo source: Yale University.

右图：迁建后的万德门，尽管屋顶上的圆亭已经没有了，但台阶加高了，看上去更雄伟壮观，气势非凡，是华西坝上老建筑门楼里最具特色的。戚亚男摄于2003年。

The Vandeman Memorial Hall after the relocation. Although the round pavilion on the roof was gone, but the steps were raised and look more magnificent and imposing. It is the most distinctive one in the gatehouse of old building on Huaxiba. Photo by Qi Yanan in 2003.

the city, and for the road construction it must be demolished. Chengdu cultural celebrity Liu Shahe wrote, "In the early 1950s, the novelist Li Jieren, then vice mayor of Chengdu, was in charge of urban construction. One day, he went to Buhou Jie No.2 yard (the location of Sichuan Provincial Federation of Literary and Art circles), and accidentally talked about the street construction planning of this city, with great interest and loud voice. He said to demolish from Gongyuan Jie (Imperial City flatland) to the south. Sanqiao Beijie, Sanqiao Nanjie, Hongzhaobi, Chunhua Jie all have to be swept away. Further south, to demolish the city wall, to cross the Jinjiang River, to demolish all the way south. Then, a road dozens of meters wide will come which will be as the central axis of the new Chengdu development to the south. Li Jieren excitedly said, 'At that time, when you ascend the Imperial City and look southward, you will see the wide ten-mile long street. That will be truly magnificent!'…You see, a straight line to the south, really cut cake forthright. Who thought of this cut, the Huaxiba in the south outside the city will be forever cut in half…Within a few years, the demolition plan became a reality and was executed. Hence there are the Tianfu Square, and the South Renmin Road. Fortunately, the cake is still a cake even though it has been cut in two pieces. Occasionally into the Huaxiba a tour, as if the scenery is still the same."

的确如流沙河所说的那样，人民南路笔直地把华西坝分为了两半。幸运的是当时利用拆下来的万德堂的旧建材在钟楼旁重新复原修建该楼，基本还其本来面目，但与原建筑不同的是屋顶上中式攒尖顶圆亭没有了。2000年，学校又在钟楼旁与万德堂对称的地方修建了第十教学楼，该楼完全模仿复原万德堂形式，这样在钟楼两旁就有两座外形一样的建筑。2013年5月，学校命名第十教学楼为仁德堂。

As Mr. Liu Shahe said, the South Renmin Road straightly divides Huaxiba into two parts. Fortunately, the Vandeman Memorial Hall has been rebuilt by using the old building materials of the demolished. The restored Vandeman Memorial Hall located next to the clock tower and the building was basically restored to its original appearance, but different from the original building, for the round pavilion with Chinese spire roof was missing. In 2000, the 10th Teaching Building was built on the place symmetrical to the Vandeman Memorial Hall beside the clock tower. The building completely imitated the restored the Vandeman Memorial Hall, so that on the east and west sides of the clock tower there are two buildings of the same shape.

1945年，成都金陵女子文理学院（位于万德堂）宽阔的阶梯上，下课的女生三三两两相伴而行，她们身着春装，手拿课本或书包，迎着初春的太阳，对着镜头开心地笑了。图片来源：耶鲁大学。

In 1945, on the broad steps of Chengdu Ginling College (located in Vandeman Memorial Hall), the girls who finished class walked together in twos and threes. They were dressed in spring clothes and holding textbooks or schoolbags, and smiled happily at the camera in the face of the early spring sun. Photo source: Yale University.

"一代的崇高女性"教育家吴贻芳
A Noble Women of Her Generation Educator Wu Yifang

1938年1月，我国教育史上第一位大学女校长、金陵女子文理学院校长吴贻芳带领该校师生从南京迁到成都，在华西坝上开始了长达8年的办学工作，直到1946年才返迁回南京。华西协合大学把万德堂的大部分房间腾出来作为金陵女子文理学院教学和行政用房，校长吴贻芳女士在该楼办公。

In January, 1938, Wu Yifang, the first female university president of China and President of Ginling College, led the teachers and students of the college to move from Nanjing to Chengdu and began to run the school on Huaxiba for 8 years. It was not until 1946 that they returned to Nanjing. West China Union University vacated most of the room in the Vandeman Memorial Hall as the Teaching and Administrative Room of Ginling College. The President, Ms. Wu Yifang, worked in this building.

吴贻芳1893年出生于江苏泰兴，16岁那年，吴贻芳的家庭发生了重大的变故，连着两三年间父亲、母亲、哥哥、姐姐

都先后离开人世，只留下了她和妹妹。姨父收养了她们姐妹俩。1915年，美国教会在南京开办了金陵女子大学，第二年吴贻芳以优异的成绩插班进入该校。1919年，吴贻芳和另外4位女同学毕业了，成为中国第一批获得学士学位的女大学生。1922年，吴贻芳获得巴勃尔奖学金并去美国密执安大学留学。1928年，吴贻芳获得生物学博士学位后回国接任金陵女子大学校长一职。两年后，在吴贻芳的领导下，金陵女子大学通过国民政府立案，学校改名为金陵女子文理学院，开始了她执掌该校23年的校长生涯。

Wu Yifang was born in 1893 in Taixing, Jiangsu Province. At the age of 16, Wu Yifang's family suffered a disaster. Her father, mother, brother and sister all passed away in two or three years, leaving only her and her sister. Her uncle adopted the two sisters. In 1915, an American church opened Ginling College in Nanjing. The next year, Wu Yifang entered the school with excellent scores. In 1919, Wu Yifang and four other female students graduated, becoming one of the first batch of female college students in China to receive a bachelor's degrees. In 1922, Wu Yifang won the Barbour scholarship to study at the University of Michigan USA. In 1928, Wu Yifang returned to China to take over as president of Ginling College after receiving her doctorate in biology. Two years later, under the leadership of Dr. Wu Yifang, Ginling College successfully registered with the Government of the Republic of China, beginning her 23-year presidency of the university.

1937年7月7日，全面抗日战争爆发，日军逼近南京，金陵女子文理学院决定以上海、武汉和成都三地为办学地点。南京沦陷后，吴贻芳决定将学校全部迁到成都。华西协合大学不仅提供了万德堂的大部分房间作为金陵女子文理学院教学和行政用房，而且还借了一块土地让金陵女子文理学院建了一栋简易的学生宿舍。吴贻芳婉言谢绝华西坝外籍教师邀请她到小洋楼去居住的建议，而是选择与学生同住简易宿舍。1938年6月，有5名学生在成都顺利毕业了；9月，金陵女子文理学院在成都恢复招生。

On July 7,1937, when the Counter-Japanese War broke out in an all-round way and the Japanese army approached Nanjing, Ginling College moved respectively to Shanghai, Wuhan and Chengdu. After the fall of Nanjing, Dr. Wu Yifang decided to move the college completely to Chengdu. West China Union University not only provided most of the room in the Vandeman Memorial Hall as the Teaching and Administrative Room of Ginling College, but also lent a piece of land to Ginling College to built a simple student dormitory on it. Dr. Wu Yifang politely declined the advice of the foreign teachers of West China Union University to invite her to live in a small house, she chose to live simple dormitory with students. In June, 1938, five students successfully graduated in Chengdu. In September, 1938, Ginling College resumed enrollment in Chengdu.

金陵女子文理学院规定不招收已婚女学生，已经入校的学生如果在读书期间结了婚就不能再留校读书。然而在成都华西坝，吴贻芳却打破了这一规定，她让已经结婚离开学校且带有一个孩子的景荷荪复学。景荷荪是战前考入金陵女子文

这张1945年2月拍摄的万德堂旁边操场的照片是华西坝的一道风景，叫"对牛弹琴"。但是它没有这个词语原来的意思，而是因为当年华西协合大学理学院生物系培育的提高产奶量的奶牛就养殖在华西坝上，坝上的音乐琴房也在附近，每当学生在琴房练琴，琴声就传到了附近正在放养奶牛的操场上，因而被五大学的学生戏称为"对牛弹琴"。图片来源：耶鲁大学。

This picture of the playground next to the Vanademan Memorial Hall taken in February 1945 is a scenery of Huaxiba, called "cast pearls before swine". But it does not have the original meaning of the idiom. This is because the Department of Biology at the School of Science of West China Union University raised dairy cows to improve milk production in Huaxiba, and the piano house of the university was also nearby. Whenever students practiced the piano in the piano room, the sound of the piano reached the nearby playground where cows were grazing, so it is joked "cast pearls before swine" by the students of the Associated Universities in Chengtu. Photo source: Yale University.

理学院的学生，在读书期间她与从法国留学回国的谢承瑞相识、相爱。1935年，景荷荪嫁给了谢承瑞，按学校的规定，景荷荪只有离开学校。第二年8月，他们俩的女儿出生了。全面抗战爆发时，谢承瑞任黄埔军校教导总队第一旅第二团上校团长，在南京保卫战中为国捐躯。失去了家园、丈夫的景荷荪带着周岁的女儿辗转流亡到成都，得知金陵女子文理学院已经在华西坝恢复了办学，她很想复学。当吴贻芳了解到景荷荪的经历后，她同意景荷荪复学，并跟同学们说，景荷荪的丈夫为国捐躯了，她是烈士的亲属，她的小女儿是烈士遗孤，我们每个人都应该关心和照顾她们母女，我们应该帮助景荷荪完成学业，并帮助她把孩子抚养教育成人。

Ginling College does not recruit married female students, and students who have been enrolled in school can not stay in school if they get married during their study. However, in Huaxiba, Chengdu, Dr. Wu Yifang broke the rule by allowing Ms. Jing Hesun, who had been married and left school and had one child, to return to school. Ms. Jing Hesun was a student of Ginling College before the war. During her study, she met and fell in love with Mr. Xie Chengrui, who returned from studying in France. In 1935, Ms. Jing Hesun married Mr. Xie Chengrui. She had to leave the school according to the school regulations. The following August, their daughter was born. When the Counter-Japanese War broke out in an all-round way, Mr. Xie Chengrui served as colonel of the second regiment of the first brigede of Repubic of China Miliary Academy and died for the country in the Battle of Nanjing. Losing her home and husband, Ms. Jing Hesun went in exile to Chengdu with one-year-old daughter. When she learned that Ginling College had resumed its school in Huaxiba, she wanted to go back to school. When Wu Yifang learned about Jing Hesun's experience, she agreed Jing Hesun to go back to school. She told the students, Ms. Jing Hesun's husband died for the country, She is the relative of the martyr, and her little daughter is martyr orphans. Each of us has the duty to take care of them. We should help Jing Hesun to complete studies, and help her to bring up her child.

1939年，宋美龄来到华西坝看望金陵女子文理学院的师生们，并在万德堂楼前与师生们合影留念。当她看到学生们都使用很粗糙的饭碗吃饭时，她对吴贻芳说："吴校长，学生怎么能用这么粗的饭碗？"接着她又说："虽是困难时期，我们的女孩子还是可以用上更好一些的碗的。"后来宋美龄专门给每位女生买了一个比较细的陶瓷碗。

In 1939, Ms. Soong Meiling came to Huaxiba to visit the teachers and students of Ginling College, and took a photo with the teachers and students in front of the Vandeman Memorial Hall. When she saw the students used very rough rice bowls for dinner, she said to Dr. Wu Yifang, "President Wu, how can students use such rough rice bowls?" Then she said, "Although it is difficult time, our girls can still use better bowls." Later, Ms. Soong Meiling specially bought a relatively fine ceramic bowl for each girl.

吴贻芳与宋美龄关系不一般，但是为了学校的发展，吴贻芳可以对宋美龄说："不！"1945年4月，吴贻芳回成都路经重庆时，宋美龄想请吴贻芳来当国民政府教育部部长，吴贻芳一口拒绝了。

It can be seen that the relationship between Dr. Wu Yifang and Ms. Soong Meiling is unusual. But for the development of the school, Dr. Wu Yifang can say "No" to Ms. Soong Meiling. On her way back to Chengdu in April, 1945, Dr. Wu Yifang passed through Chongqing, where Ms. Soong Meiling invited Dr. Wu Yifang to be the country's Education Minister. She refused.

吴贻芳就是这样一位女性，为了事业她终生未婚。作家冰心称吴贻芳为"一代的崇高女性"，她说："我没有当过吴贻芳先生的学生，但在我的心灵深处总是供奉着我敬佩的老师——吴贻芳先生。"

Dr. Wu Yifang is such a woman who for her career never married. Writer Bingxin called Wu Yifang "a noble women of her generation", and said, "I have never been a student of Mr. Wu Yifang, but in the depths of my heart I always worship Mr. Wu Yifang, the teacher I admire."

金陵女子文理学院的乡村服务处
Rural Services Office of Ginling College

社会学系是金陵女子文理学院最具特色的品牌专业，该系所讲授的课程着重于社会服务、社会工作及社会调查，侧重于实用方面。吴贻芳把该系作为培养社会服务人才的重要阵地。学校迁到成都以后，为了教学的需要，金陵女子文理学院在仁寿县建立了一个乡村服务处，以利于学生实习。乡村服务处结合四川当时的具体社会情况，开办了夏令营儿童校、婴儿班、小先生训练班、母亲会、妇女夜校、挑花组等。

The Department of Sociology is the most distinctive brand major of Ginling College. The courses taught by the department focus on social service, social work and social investigation, and emphasis practical aspects. Dr. Wu Yifang regarded the department as the goal of training social service talents. After the school moved to Chengdu, for the needs of teaching, Ginling College established a Rural Service Office in Renshou County conducive students to practice. Combined with the specific social conditions of Sichuan at that time, the service office carried out summer camp children's school, baby class, little teacher training, mothers' association, women's evening school, cross-stitch group and so on.

夏令营儿童校是专门为学龄儿童开办的，除了教学龄儿童识字、算术、常识外，还要教学龄儿童唱歌、做游戏。

Summer Camp Children's School was specially opened for school-age children, it supplied school-age children courses like literacy, arithmetic, common sense, also singing and gaming.

婴儿班是为3~5岁的儿童开设的，每天有一个小时的时间，除了教他们唱歌、做游戏、讲故事外，主要培养他们良好的卫生习惯以及礼貌待人的习惯。

The Baby Class was for children aged 3 to 5 years old for an hour a day. Besides singing, playing games and telling stories, it mainly cultivated them with good hygiene and politeness.

小先生训练班是培养当地的华英小学的高年级学生能够胜任夏令营儿童校和婴儿班的教学工作，为当地培养一批能从事幼儿工作的老师。

Little Teacher Training was to train the senior students of local Huaying Primary School to be competent for the teaching work of children's school and baby class in the summer camp, and to train a group of local teachers for babies and school age children.

母亲会是组织婴儿班学生的妈妈、姐姐、奶奶或是家庭里的其他女亲属，每周聚会一次，目的是既联络她们的感情，同时又向她们灌输育儿知识。每次聚会时金陵女子文理学院的学生都要带一些西红柿分给来参加活动的妈妈们吃，妈妈们对西红柿既陌生又好奇，因为之前四川地区种植西红柿的地方很少。一开始，妈妈们还很难接受，一些人不敢吃，一些胆大的接过来咬了一口，剩下的就扔掉了。后来，带去的西红柿不但被妈妈们分吃光了，她们还向大学生要西红柿的种子，想要自己种植西红柿，这无形中还起到了推广种植西红柿的作用。

Mothers' Association organized the mothers, sisters, grandmothers or other female relatives in the family of the baby class, to party once a week to understand their feelings and to tell them parenting knowledge. At each party, students from Ginling College took some tomatoes to the mothers who participated in the activity. The mothers were strange and curious about tomatoes, for tomatoes were planted little in Sichuan before. At first, it was hard for the mothers to accept, some were afraid to eat, some were bold enough to take a bite, and the rest were thrown away. Later, the tomatoes were not only eaten up by mothers, mothers also asked college students for the seeds of tomatoes to grow their own tomatoes, virtually also played a role in promoting the tomatoes planting.

乡村服务处的学生们了解到当地妇女农闲时爱做挑花手工，她们使用崇州出产的夏布用不脱色丝线挑花做成手工艺品，这些手工艺品可以用来装饰居家环境。学生们觉得这些手工艺品如果出售能给妇女们带来一些经济收入，因而他们就组织当地喜欢挑花的妇女，"绣"一些中国传统图案在桌布、床单、窗帘、餐巾上，由学校出面将这些手工艺品送到美国去出售，这样可以给她们的家庭增加一些收入，改善她们的生活。

Students of the Rural Service Office came to know that local women love to do cross-stitch handwork. They used the grass cloth of Chongzhou to do cross-stitch with colorfast silk thread and used these handicrafts to decorate their home. Students thought if these handicrafts were sold, they could bring some economic income to women. So the students organized the local women who like to do cross-stitch handwork, to embroider some Chinese traditional patterns on table cloth, bed sheets, curtains, napkins, and by the school to send to the United States to sell, so that the local women could increase their families some income, and improve their lives.

所以说，乡村服务处既让学生们在进入社会前就先深入到社会里，认识了社会，了解到社会的需求，同时又解决了当地民众的一些问题，提高了当地民众的素质，增加了他们的收入。

Therefore, the Rural Service Office not only made the students to go deep into the society before entering the society, to understand the society and the needs of the society, but also solved some problems of the local people in advance and improved the quality of the local people, and increased their income.

华西坝老建筑的前世今生
The Stories of the Historic Architecture of Huaxiba

拾

汇聚众多中国文化学者的文学院

广益大学舍

The Friends' College Building, the Liberal Art School that Gathers Many Chinese Scholars

广益大学舍又名稚德堂，1925年竣工，由英国公谊会亚兴氏所捐建。该学舍大楼平面为"H"形，屋顶是一个横向歇山顶与两侧的两个歇山顶相交的组合体，屋顶下的重檐是一楼的屋檐，入口为中式木牌坊门楼。屋顶鱼尾脊兽张口露牙咬住屋脊镇宅保平安，可见建筑师完全了解中国传统建筑中脊兽的作用。该学舍大楼原是文学院中国文学系教学楼及学生宿舍，1952年全国高等学校院系调整，中国文学系调整到国内其他院校，广益大学舍先后作为大学附属的中学、小学、幼儿园校舍，现为四川大学华西幼儿园。

The Friends' College Building was completed in 1925, donated by the Arthington Estate of British Friends Mission Association. The building plane is "H-shaped". The roof is a combination of the broadwise saddle roof with the two saddle roof of two flanks, and the double eaves roof under the roof are the eaves of the first floor, and the entrance is the Chinese wooden archway gatehouse. The fish tail ridge beast opens its teeth biting the roof to keep the building safe, which shows that the architect fully understand the role of ridge animals in traditional Chinese architecture. The Friends' College Building was originally the teaching building and student dormitory of the Chinese Literature Department of College of Arts. In 1952, there was an adjustment of the departments of colleges and universities nationwide, the Department of Chinese Literature was adjusted to other universities in China. Then, the Friends' College Building had been

广益大学舍楼前的广场上竖起简易的篮球架就是一个很好的篮球场，大楼右边带5层塔楼的建筑是1914年修建的亚克门纪念室。图片来源：四川大学档案馆。

A simple basketball rack erected in the square in front of the Friends' College Building is a good basketball court. On the right the building with the 5-story tower is the Ackerman Memorial Dormitory built in 1914. Photo source: Sichuan University Archives.

1952年之后，广益大学舍先后作为大学附属的小学、中学和幼儿园校舍。20世纪80年代初，为了拓展教室空间，拆除了大楼四周的台基并且给底层安装窗户用作教室使用，同时在台阶两旁增加栏杆。现在该大楼是学校的附属幼儿园。咸亚男摄于2011年。

Since 1952, the Friends' College Building had successively been served as the primary school, middle school and kindergarten affiliated to the university. In the early 1980s, in order to expand the classroom space, the platform base around the building was removed and windows were installed on the ground floor for classrooms use. Meanwhile, railings were added on both sides of the steps. The building is now a kindergarten attached to the university. Photo by Qi Yanan in 2011.

the building of a secondary school and a primary school affiliated to the university, and is now Huaxi Kindergarten Building of Sichuan University.

抗日战争时期，华西坝汇聚了国内众多的学者名流，如陈寅恪、顾颉刚、钱穆、闻宥、林山腴、许寿裳、沈祖棻、吕叔湘、吴宓……他们都在广益大学舍讲台上留下了身影。而在1940年成立的中国文化研究所就设立在广益大学舍，该所发行的中外文学术刊物《华西协合大学中国文化研究所集刊》曾蜚声海内外。

During the Counter-Japanese War, many domestic scholars and celebrities gathered in Huaxiba. They were Mr. Chen Yinke, Mr. Gu Jiegang, Mr. Qian Mu, Mr. Wen You, Mr. Lin Shanyu, Mr. Xu Shoushang, Ms. Shen Zufen, Mr. Lü Shuxiang, Mr. Wu Mi… They all had taught in the Friends' College Building. In particular, the Institute of Chinese Culture was founded in 1940 in the Friends' College Building. The academic journal of the Institute named *Collection of the Institute of Chinese Culture of West China Union University* was well known both at home and abroad. The jounal was published in Chinese and foreign languages.

2019年1月，广益大学舍大楼被列入成都市历史建筑保护名录。

In January, 2019, the Friends' College Building was listed on the Historic Buildings Protection List of Chengdu.

中国文学系的创办者程芝轩
Cheng Zhixuan, Founder of the Department of Chinese Literature

1910年,华西协合大学创办之初就开设了文科和理科,仰慕博大精深的中国文化的外国传教士们不但给自己起了一个中国名字,而且还聘请中国文化大师来教会大学讲授中国文化。他们说:"东西方结合起来,发挥自己的长处,汲取精华,满足中国新的一天新的需要。"

In 1910, at the beginning of its establishment, West China Union University offered liberal arts and science courses. Foreign missionaries who admired the profound Chinese culture not only gave themselves Chinese names, but also hired Masters of Chinese culture to teach Chinese culture in missionary universities. They said, "The East and the West come together to play their strengths and absorb the essence to meet China's new needs for a new day."

当时在五四反帝爱国运动的影响下,一场声势浩大的收回教育主权的运动在全国展开了。为了使学校能继续办下去,华西协合大学校方做出了决定:第一,在政府立案;第二,服从政府条件;第三,着重中国文学及历史教学;第四,稳步增加中国人员在教职员及理事部中的比例。

At that time, under the influence of the "May 4th" Anti Imperialist Patriotic Movement, a huge movement to recover the sovereignty of education was launched in the whole country. In order to enable the university to continue, the university made the following decisions. First, register with the government; Second, obey the government conditions; Third, focus on the teaching of Chinese literature and history; Fourth, steadily increase the proportion of Chinese personnel in the faculty and council.

1924年,校方聘请前四川巡按使公署教育科长程芝轩来校任职,为学校的生存和发展出谋划策。程芝轩刚来华西时任总务部主任,全面负责校务工作。为了尽早成立中国文学系,程芝轩兼任图书馆中文部主任。当时图书馆的中文图书很少,不利于开展中国文学的教学,他四处搜集中文经典图书。其后校方获得美国哈氏奖学基金,专门用于筹建中国文学系,程芝轩用此款购买了大量的中文图书,并且又特聘民国初年蜚声成都的"五老七贤"来校任文科教授。

In 1924, the university invited Mr. Cheng Zhixuan, the former Education Section Chief of Sichuan Province to take a post in the university to give advice for the survival and development of the university. Mr. Cheng Zhixuan was the Director of the General Affairs Department and was fully in charge of the school affairs work when he just came to West China Union University. In order to establish the Department of Chinese Literature as soon as possible, Mr. Cheng Zhixuan also served as the Director of the Chinese Department of the Library. At that time, there were few Chinese books in the library, which was not conducive to the teaching of Chinese literature. He

1931年6月，华西协合大学中国文学系高明秦、洪有模两位毕业生（前排）与老师、同学在广益大学舍门楼前合影留念。照片里第二排左边第二位是文学系主任程芝轩，第三排左边第二位是教务长方叔轩。图片来源：四川大学档案馆。

In June, 1931, Gao Mingqin and Hong Youmo, two graduates of the Department of Chinese Literature of West China Union University (front row), took a photo with their teachers and classmates in front of the Friends' College Building. The second place on the left of the second row is the Dean of Literature Cheng Zhixuan. The second place on the left of the third row is the Provost Fang Shuxuan. Photo source: Sichuan University Archive.

collected classic Chinese books everywhere. Later, the university obtained the Harvard Fund from the United States, which was specially used for the establishment of the Department of Chinese Literature. Mr. Cheng Zhixuan bought a large number of Chinese books with the fund, and hired the "Five Old and Seven Sages", the representative figures of Sichuan Confucianism in the late Qing Dynasty and the early Republic of China to be professors of liberal arts at the university.

1927年，中国文学系宣告成立，程芝轩出任主任，并开始招生。由于华西协合大学在中国文化事业上已经做出了成绩，1928年哈佛燕京学社成立后，提供了20万美元用于支持华西协合大学的中国文学系、图书馆和古物博物馆的发展。

In 1927, the Department of Chinese Literature of West China Union University was established, Mr. Cheng Zhixuan as director, and began to recruit students. Since West China Union University had made great achievements in Chinese cultural affairs, after the establishment of Harvard-Yenching Institute in 1928, it provided 200,000 US dollars to support the development of the Department of Chinese Literature, Library and Antiquities Museum of West China Union University.

1933年，程芝轩又创办中国文学学术刊物《华西学报》，专门刊载国学论著以及文学作品等，该刊物"取材编辑，俱极慎重，内容尤为充实，发行以来深得学术界之推重"。

In 1933, Mr. Cheng Zhixuan founded the Chinese Literature Academic Journal *West China Journal*, which specialized in sinology works and literary works. The journal "selected materials and edited very carefully. The content is particularly rich. Since the release, the journal has been highly valued by the academic community".

中国文学系成立以后，程芝轩又忙于为学校向政府立案的工作。程芝轩清末民初曾在四川教育界工作过，他不仅与时任四川省公署教育厅厅长沈与白有交情，而且与南京国民政府教育部执政当局官员也多有交往，因而他是立案工作的具体经办者。他频繁来往于学校与政府部门之间，传递信息与沟通。立案之事经过6年的艰苦努力，1933年教育部下令："私立华西协合大学，应准予立案"。至此，华西协合大学成为中国政府管理下的一所私立大学。

After the establishment of the Department of Chinese Literature, Mr. Cheng zhixuan was busy filing cases with the government for the university. Mr. Cheng Zhixuan was in charge of education in Sichuan at the end of Qing Dynasty and the beginning of ROC. He not only had good relations with Shen Yubai, the Director of Education Department of Sichuan Provincial Government, but also had many contacts with the officials of Ministry of Education of Nanjing National Government. So he was the specific handler of case filing work. He frequently went back and forth between the university and the government to pass on information and communication. After six years of hard work, in 1933, the Ministry of Education ordered that "the private West China Union University should be allowed to register". From then on, West China Union University became a private university under the administration of the Chinese government.

蜚声海内外的中国文化研究所
Renowned Institute of Chinese Culture at Home and Abroad

1939年，华西协合大学得到了哈佛燕京学社资助，开始筹建中国文化研究所，专门研究中国的宗教学、考古学、史学、人类学、语言学等学科。精通多国语言文字，国学水平很高的闻宥被华西协合大学选中，想让他担任研究所所长。但当时闻宥在昆明国立云南大学任职，为此教务长方叔轩专程到昆明邀请闻宥出任研究所所长，任教授兼中文系主任。

In 1939, West China Union University was funded by the Harvard-Yenching Institute for establishment the Institute of Chinese Culture, specializing in Chinese religion, archaeology, historiography, anthropology, linguistics and other disciplines. Mr. Wen You, who was proficient in several languages and had a high level of sinology, was selected as the director of the Institute by West China Union University. But at that time, Mr. Wen You served National Yunnan University in Kunming, so dean Fang Shuxuan (S. H. Fong) made a special trip to Kunming to invite Mr. Wen You as the Director of the Research Institute, professor and Director of the Chinese Department.

闻宥来到华西坝后，充分利用抗战期间国内许多知名教授和学者都云集在西南后方的条件，他先后聘请陈寅恪、刘咸、韩儒林、李方桂、董作宾、吴定良、吕叔湘、缪钺等人，甚至还有德国汉学家傅吾康、法国藏学专家石泰安来研究所从事研究工作。1940年9月，一本采用横排（当时的中文书刊都是竖排）、中文外文并存、面向国际的、以研究中国文化为主的学术期刊问世了，该刊的中文名称为《华西协合大学中国文化研究所集刊》，外文名称为 STUDIA SERICA。该刊所刊登的文章主要采用该研究所学者的论文，也适当刊登非该研究所学者的文章。该刊最特别的是学者撰写的文章不以中文为限，英、法、德文皆可使用，而且要求用中文写的需附外文提要，用西文写的需附中文提要。其目的就是要让该刊成为一本国际化的学术刊物，让中外学者都能更好地阅读。该刊物每年出一卷，一卷又分几号分别出版发行。

语言学大师闻宥。图片来源：《持志年刊》1928年第3期。
Language master Wen You. Photo source: *The Chih Tze Annual* No. 3, 1928.

After Mr. Wen You came to West China Union University, he made full use of the conditions that many well-known professors and scholars gathered in the southwest rear during the Counter-Japanese War. He hired Mr. Chen Yinke, Mr. Liu Xian, Mr. Han Rulin, Mr. Li Fanggui, Mr. Dong Zuobin, Mr. Wu Dingliang, Mr. Lü Shuxiang, Mr. Miao Yue, and even German sinologist Mr. Wolfgang Franke, French Tibetan experts Mr. Rolf Alfred Stein to institute engaged in research work. In September, 1940, a horizontal row (at that time, Chinese books and periodicals are vertical row), Bilingual Chinese and foreign language, internationally-oriented academic journal mainly studying Chinese culture was published. The Chinese name of the journal was *Collection of the Institute of Chinese Culture of West China Union University*. The foreign language name was *STUDIA SERICA*. The articles published in this journal were mainly written by scholars in the institute, but also those of non-institute scholars. The most special feature of this journal was that the articles written by scholars were not limited to in Chinese, but also could be in English, French and German. It was also required that those written in Chinese should attach a foreign language abstract and those written in Western should attach a Chinese abstract. The aim was to make the journal an international academic journal, so that both Chinese and foreign scholars could better read. The publication was published in one volume annually and in several installment.

尽管有哈佛燕京学社的经费支持，但在抗战最艰苦的年代要想出版这样一本学术刊物还是很不容易的。为了使刊物能够尽快出版发行并在第一时间分发到西方一些相关的学术机构，以便更快地得到反馈，闻宥把该刊物放在上海出版印刷。研究所委托人把手稿送到上海，并请在上海的朋友校稿，因而该刊物在研究所成立的当年就出版发行了。很快就有国内外的大学、图书馆和学术机构汇款前来订购。

Despite the financial support of the Harvard-Yenching Institute, it was not easy to publish such an academic publication in the hardest era of the Counter-Japanese War. In order to publish the first issue as soon as possible and distribute it to some relevant western academic institutions in the first time, so as to get faster feedback, Mr. Wen You put the first issue in Shanghai for publishing and printing. The institute entrust people to send the manuscript to Shanghai and invited friends in Shanghai to proofread it. So the first issue was published and issued in the year when the institute was established. Soon universities, libraries and academic institutions at home and abroad sent remittance to order the journal.

闻宥在华西坝除了担任中国文化研究所所长从事学术研究外，还在中文系任教授，给学生上课，他开设"文字学""语言学"和"声韵学"等课程。闻宥教的学生除了有本国的，还有外国的，其中最为大家熟知的是诺贝尔文学奖18位终身评委之一的瑞典著名汉学家马悦然。1948年，马悦然获得美国洛克菲勒基金会的奖学金，到四川来进行方言调查。由于闻宥主编的《华西协合大学中国文化研究所集刊》在国际上享有很高的声望，以及其本人在语言学上的造诣，因而马悦然很自然地就找到闻宥求助，闻宥在广益大学舍收下了这位洋学生。不久马悦然就写出了他的第一篇方言学论文，并且发表在《华西协合大学中国文化研究所集刊》上。多年以后，马悦然回忆道："他当时的年龄足以做我的父亲，而从我们友谊的最初开始，他对于当时在学术道路上艰难探索的我来说也确像一个父亲。他对我在研究四川方言语音、音韵上给予的耐心帮助和智慧启迪让我永远感念。"

In Huaxiba, in addition to serving as the director of the Institute of Chinese Culture, Mr. Wen You also was a professor in the Chinese Department, to teach students. He opened Philology, Linguistics, Phonology and other courses. Not only there were domestic students, but also foreign students, among them the best known was the famous Swedish Sinologist Goeran Malmqvist, who is now one of the 18 lifetime judges of the Nobel Prize for Literature. In 1948, Goeran Malmqvist received a fellowship from the American Rockefeller Foundation and came to Sichuan to conduct a dialect investigation. Due to the international reputation of the journal, *Collection of the Institute of Chinese Culture of West China Union University* edited by Wen You, as well as Wen You himself in the attainments of linguistics, Goeran Malmqvist naturally found Mr. Wen You for help. Mr. Wen You accepted this foreign student in the Friends' College Building. Soon Goeran Malmqvist wrote his first dialect paper and published it in the *Collection of the Institute of Chinese Culture of West China Union University*. Many years later, Goeran Malmqvist recalled, "he was old enough to be my father, and from the very beginning of our friendship, he was a father figure to me as I struggled along the academic path. I will always be grateful for his patient help and intellectual enlightenment to me in studying the pronunciation and phonology of Sichuan dialect."

闻宥主持的中国文化研究所出版的《华西协合大学中国文化研究所集刊》在1950年出版了第9卷后就停刊了。随着20世纪50年代初全国高校院系调整，华西协合大学调整为四川医学院，其文、理学科被调整到其他院校，学校由一所综合性大学转变成了医药学院，闻宥也被调整到四川大学。调整后华西坝就没有中国文化研究所了。

　　The collection of the Institute of Chinese Culture of West China Union University published by the Institute of Chinese Culture chaired by Mr. Wen You, was stopped after volume 9 was published in 1950. With the adjustment of the departments of colleges and universities nationwide in the 1950s, West China Union University was adjusted from a comprehensive university to a college of medicine, renamed Sichuan Medical College. Its liberal arts and science disciplines were adjusted to other colleges and universities, and Mr. Wen You was also transferred to Sichuan University. There is no the Institute of Chinese Culture in Huaxiba after the adjustment.

这是1948年华西协合大学方叔轩校长写给闻宥的信，2010年由闻宥的儿子闻广提供给笔者。

信中写道："安德森先生和马悦然先生刚到华西坝，他们现在住在米玉士博士家里，他们想尽快会见你，讨论他们的学习计划。请你告诉我你什么时间能见他们，以便我告诉他们什么时候去广益大学舍拜访你。"

This letter was written to Wen You by S.H.Fong, the president of West China Union University in 1948, and provided to author by Wen Guang, the son of Wen You in 2010.

The letter said,"Dear Prof. Wen, Mr. Olle B. Anderson and Mr. Goeran Malmqvist have just arrived. They would like to meet you as soon as possible to talk over their plan of study. Will you kindly let me know when you can see them, so that I may inform them when to call upon you at Friends College. They are staying with Dr. and Mrs. Meuser."

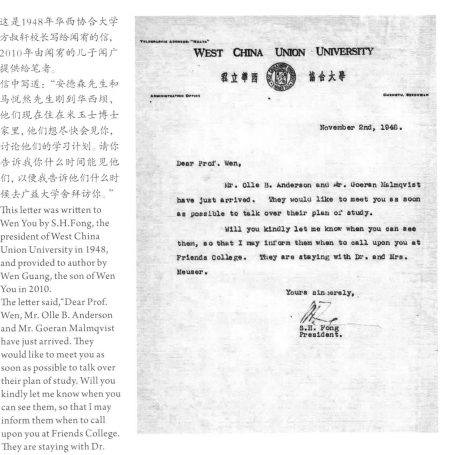

国学泰斗陈寅恪在广益大学舍开讲
The Master of Sinology Chen Yinke Lectured at Friends College

陈寅恪是中国史学界的领军人物,抗日战争前,陈寅恪是清华大学汉学研究所的一名教授。1937年七七事变后,国立北京大学、国立清华大学和私立南开大学南迁到昆明组成国立西南联合大学继续办学,陈寅恪带领家人随校南迁。1937年11月3日,为了躲避日本人,陈寅恪一家人悄悄地离开了北平,这使陈寅恪没有机会去治疗眼疾,导致其右眼失明。第二年年初,他们一家人才辗转来到香港,安顿好家人,陈寅恪便赶赴昆明西南联合大学任教,每到假期他就回到香港与家人团聚。

Mr. Chen Yinke is the leading figure in Chinese historiography. Before Counter-Japanese War, he was a professor of the Institute of Sinology of Tsinghua University, After the July 7th Incident of 1937, National Peking University, National Tsinghua University and Private Nankai University moved south to Kunming to form the National Southwest United University to continue running schools. Mr. Chen Yinke led his family to move south with the University. On November 3, in order to avoid the Japanese, he had no chance to treat his eye diseases, resulting in blindness in his right eye, and the family left Beiping quietly. At the beginning of the next year, Mr. Chen Yinke and his family came to Hong Kong to settle his family. Mr. Chen Yinke rushed to Kunming National Southwest Union University to teach. He returned to Hong Kong to reunite with his family every holidays.

1941年12月7日太平洋战争爆发。同年12月26日,日军占领了香港,陈寅恪一家人暂时安定的生活又被打破了。很快就有人登门请陈寅恪到沦陷的广州或是上海、北平任教,为了再次躲避日本人,陈寅恪又带领家人于1942年5月5日逃离了香港。1943年12月,陈寅恪一家人流亡到成都,住在陕西街,此次陈寅恪来成都是应燕京大学在成都复校之聘。3年的流亡生活,陈寅恪除了损失他写的大量书稿外,他的身体健康状况也越来越差。时任华西协合大学中国文学系主任、中国文化研究所所长的闻宥邀请陈寅恪来该所任特约研究员,一来陈寅恪可以多一份收入,二来他也可以搬到华西坝来居住,这样对他的健康和做学问都有好处。

On December 7,1941, the Pacific War broke out. On December 26, 1941, the Japanese army occupied Hong Kong, and the temporary stable life of Mr. Chen Yinke's family was broken again. Soon, he was asked to teach in the occupied Guangzhou, or Shanghai and Beiping. In order to escape the Japanese again, Mr. Chen Yinke led his family to flee Hong Kong on May 5,1942. In December, 1943, he and his family went to Chengdu in exile and lived at Shanxi Street. He came to Chengdu and was hired by Yenching University to resume the school in Chengdu. After three years of living in exile, Mr. Chen Yinke lost a lot of his manuscripts he wrote, and his health deteriorated. At that time, Mr. Wen You, Director of the Department of Chinese Literature and Director of the Institute of Chinese Culture of

West China Union University, invited Mr. Chen Yinke to be a special researcher. On the one hand, Mr. Chen Yinke could get more income, and on the other hand, he could also move to Huaxiba to live, which was good for his health and doing research.

  1944年暑期，陈寅恪一家人搬到了广益大学舍后面的一座小洋楼的一楼，陈寅恪的女儿后来回忆这里"环境幽静""院门外树木葱茏""成为我们逃难以来最好的一处住宅"。陈寅恪给学生上课很方便，出门十几步就到广益大学舍。当时陈寅恪的讲课除了成都燕京大学的学生来听外还有华西坝上其他学校的学生也会来听课。一天，陈寅恪在给学生上课时，看见成都"五老七贤"之一的林思进也与同学们坐在下面，他很惊讶，说："山公厚我，励我，真我之良师也。"

  In the summer vacation of 1944, Mr. Chen Yinke and his family moved to the first floor of a small foreign style house behind the Friends' College Building. His daughter later recalled the place as "quiet and secluded environment" and "luxuriant trees outside the courtyard", which "became the best residence since we escaped". It was very convenient for Mr. Chen Yinke to give classes to his students. It was only more than ten steps away from the Friends' College. At that time, his lectures were attended by students not only from Chengdu Yenching University but also from other universities in Huaxiba. One day, when Mr. Chen Yinke was teaching his students, he saw master Lin Sijin, one of the "Five Old and Seven Sages" of Chengdu, sitting with students.He was surprised and said: "Master Lin values me and encourages me. He is my true good teacher."

  尽管陈寅恪在这人称"天堂"的华西坝从事教学和国学研究工作，但也没能让他从创伤中康复过来。1945年年初，他唯一能使用的左眼也失明了。双目失明的残酷现实一度让陈寅恪非常沮丧，他请林思进书写他集李商隐诗句"今日不知明日事，他生未卜此生休"。林思进觉得此联过于伤感，他对陈寅恪说："君自有千秋之业，何言此生休耶！"并谢以不能书。陈寅恪还是另请友人书写此联并悬挂于家中。不久陈寅恪想通了，他请四川省教育厅厅长郭有守夫人杨云慧书写他新集苏东坡诗句"闭目此生新活计，安心是药更无方"来代替前联。

  Although Mr. Chen Yinke engaged in teaching and sinology research at Huaxiba, known as "heaven", he did not recover from the trauma and lost his only available left eye vision in early 1945. Frustrated by the harsh reality of being blind, he asked master Lin Sijin to write down a poem he liked that of Tang Dynasty poet Li Shangyin: "Today I don't know what tomorrow will be. The next life is unpredictable, and this life should be no meaning". Master Lin Sijin felt that the poem was too sentimental, so he said to Mr. Chen Yinke, "You have your own career. How can you say that this life is meaningless?" and refused to write these two sentences to him. Mr. Chen Yinke still asked another friend to write down the sentences and hung it at home. Soon he figured it out. He asked the provincial education department director Guo Youshou's wife Yang Yunhui to write down a poem he liked that of poet Su Dongpo, "Close your eyes and do a new job in this life, peace of mind is the best prescriptions" to replace the former.

陈寅恪在成都失去了他唯一可用的左眼视力，成都是陈寅恪的伤心之地，但他却把成都当作了他的故乡。陈寅恪在《忆故居》里这样写道："渺渺钟声出远方，依依林影万鸦藏。一生负气成今日，四海无人对夕阳。破碎山河迎胜利，残余岁月送凄凉。松门松菊何年梦，且认他乡作故乡。"

Mr. Chen Yinke lost his only available left eye vision in Chengdu. So, Chengdu is his sad place. But he regards Chengdu as his hometown. He wrote a poem, *Recalling the Former Residence*. Here is the poem: The bell sounds distant and melodious. The shadows of the woods are lovely, and the crows return to their nests in the woods. All my life, I stick to my pursuit, and I have to stick to my independent mind alone. The broken mountains and rivers ushered in the victory, but the remaining years still could not drive away the sadness. Bamboo door, pine and chrysanthemum have become a dream, so I would like to regard the other city as my hometown.

百代签约歌手、声乐教育家郎毓秀
EMI Record Signing Singer, Vocal Educator Lang Yuxiu

郎毓秀这个名字对于20世纪三四十年代的人来说应该是不陌生的，她是与黄友葵、喻宜萱、周小燕齐名的著名女高音歌唱家。抗战期间郎毓秀流亡到成都，在成都开音乐会，之后被华西协合大学邀请担任声乐教授，在广益大学舍成立了音乐系并出任系主任一职。

The name Lang Yuxiu should be no stranger to people in the 1930s and 1940s. She was a famous soprano as famous as Ms. Huang Youkui, Ms. Yu Yixuan and Ms. Zhou Xiaoyan. During the Counter-Japanese War, Ms. Lang Yuxiu went in exile to Chengdu and held concerts in Chengdu. Later, she was invited to be a vocal professor by West China Union University. She established the Music Department in the Friends' College Building and served as the head of the department.

1918年，郎毓秀出生在上海，她的父亲郎静山是我国第一代从事摄影的大师。郎静山在家里冲洗照片时，总要放西洋音乐的唱片听，殊不知这就成了童年的郎毓秀对声乐爱好的启蒙教育。1934年，郎毓秀考入了当时国内唯一的一所高等音乐学府——上海国立音乐专科学校主修声乐，师从俄籍苏石林教授。在校时，她的歌已唱得非常好了。周小燕曾经回忆起她是如何认识比她高一年级的郎毓秀：有一天她经过教室走廊，听到一位女高音的歌声从教室里传了出来，那是何等令人羡慕的歌声啊，高亢圆润，音域好宽，音色是金属的。她一下被镇住了，不由得止住脚步侧耳细听。后来一打听，唱歌的女生叫郎毓秀，唱的是歌剧《托斯卡》的咏叹调《为艺术，为爱情》，用的是地道的意大利美声唱法。

Ms. Lang Yuxiu was born in Shanghai in 1918. Her father, Mr. Lang Jingshan, was among the first generation of masters engaged in photography in China. While her father was developing

1945年，郎毓秀在成都开演唱会。郎毓秀提供。
In 1945, Ms.Lang Yuxiu held a concert in Chengdu. Lang Yuxiu provided.

photos at home, he always listened to western music records. This became the childhood Lang Yuxiu to vocal music hobby enlighten education. In 1934, Ms. Lang Yuxiu was admitted to Shanghai National Music College, the only institution of higher music in China, majoring in vocal music under Russian Professor Su Shilin. At the college, she performed very well. Ms. Zhou Xiaoyan once recalled how she knew Ms. Lang Yuxiu, who was one year higher than her. One day she passed the classroom corridor and heard a soprano singing from the classroom. How admirable the song was, high and mellow, the sound range was so wide, and the timbre was metallic. She was suppressed, she could not help but stop listening carefully. Later an inquiry, the singing girl, named Lang Yuxiu, sang the opera *Tusca Aria to Love for Art*, in the authentic Italian bel canto.

不久，高年级的同学贺绿汀写的一首四重唱《湖堤春晓》被百代唱片公司看上，他喜欢郎毓秀的声音，就请她担任女高音录制唱片。《渔光曲》的作者任光那时正在百代公司担任经理，听到《湖堤春晓》后，想让郎毓秀做签约歌手，但她当时只有16岁。任光找到她父亲洽谈并签订了两年的合同，就这样郎毓秀成了百代公司的签约歌手。随后，她在百代公司录制了30多张唱片，如《大军进行曲》《杯酒高歌》《乡愁》《满园春色》《天伦》等等，都是黄自、刘雪庵、贺绿汀等名家作品。

Soon, senior student He Lüting wrote a quartet *Lake Embankment Spring Dawn* and was settled on by EMI Records. Mr. He Lüting liked Ms. Lang Yuxiu's voice, and asked her to serve as a soprano to record. Ren Guang, the author of *Fishing Light Song*, was a manager of EMI and wanted Ms. Lang Yuxiu to be a contract singer, but she was only 16 years old at that time. Mr. Ren Guang found her father and negotiated to sign a two-year contract, so that Ms. Lang Yuxiu became a contract singer of the EMI. Subsequently, she recorded more than 30 records in EMI, such as *March of the Army, Singing With A Glass of Wine, Nostalgia, Spring Scenery in the Garden, Tian Lun* and so on, all of which are works of famous artists Mr. Huang Zi, Mr. Liu Xuean and Mr. He Lüting.

1937年，郎毓秀听了刚从法国巴黎音乐学院留学归来的冼星海建议赴比利时布鲁塞尔皇家音乐学院主修声乐。1941年，郎毓秀毕业回国。之后，因在北平无法忍受做亡国奴的屈辱生活，她与家人离开沦陷的北平，几经周折于1945年定居成都。抗战胜利后，郎毓秀又赴美国俄亥俄州辛辛那提师范学院进修声乐。

In 1937, Ms. Lang Yuxiu accepted the suggestion of Mr. Xian Xinghai, who had just returned from the Paris Conservatory of Music, went to major in vocal music at the Royal Conservatory of Music in Brussels, Belgium. In 1941, Ms.Lang Yuxiu graduated and returned home. Because she could not bear the humiliating life of being a subjugated slave in Peking, she and her family left the occupied Peking and settled in Chengdu in 1945. After the victory of the Counter-Japanese War, Ms. Lang Yuxiu went to Cincinnati City Normal College, Ohio to study vocal music.

1948年9月，学业有成的郎毓秀婉言谢绝了美国朋友的盛情挽留，回到了成都。华西协合大学加拿大籍钢琴教授汪德光得知郎毓秀已回到成都，立刻说服校方聘请她出任音乐系主任，这样原先没有音乐系的华西协合大学，也因为郎毓秀的到来而成立了音乐系，第二年音乐系招了6位学生。

In September, 1948, the academic successful Ms. Lang Yuxiu politely declined the invitation of American friend and returned to Chengdu. Ms. A. Ward, a Canadian piano professor at the West China Union University, learned that Ms. Lang Yuxiu had returned to Chengdu and immediately persuaded the college to hire her as the dean of the Music Department. Thus, the college without the Music Department also established the Music Department because of her arrival. In the next year, the Music Department recruited six students.

1952年全国高校院系调整时，郎毓秀随华西大学音乐系调整到西南音乐专科学校（现四川音乐学院）担任声乐系任主任。尽管郎毓秀在华西协合大学时只有一位学生毕业，但来到四川音乐学院任教后，那可是桃李满天下，她90多岁高龄的时候都还在从事教学、写回忆录，她把一生都献给了钟爱的音乐。

In 1952, when the nationwide department of colleges and universities were adjusted, Ms. Lang Yuxiu with the Music Department of West China Union University was transferred to the Vocal Department of Southwest Music College (now Sichuan Conservatory of Music). Although Ms. Lang Yuxiu had only one student graduated at West China Union University, after she came to the Sichuan Conservatory of Music, her student was everywhere. In her 90s, she was still teaching and writing memoirs to devote her all to her favorite music.

华西坝老建筑的前世今生
The Stories of the Historic Architecture of Huaxiba

拾壹

中西文化的交汇处

―

合德堂

The Hart College, Where East Culture Meets West

合德堂又名赫斐楼、赫斐院，1915年动工，1920年建成，由加拿大美道会为纪念到中国西南传教的美国人赫斐秋所捐建。原为华西协合大学理学院大楼，现为成都美国海外留学中心。合德堂采用以三重檐四角攒尖顶的方形塔楼与单檐歇山顶为主的中西结合的方式建造。在中式传统大屋顶上建西式塔楼是很有创意的。合德堂的重檐是华西坝上老建筑中最多的，有五重。门楼上的方形塔楼给整座大楼增添了庄重肃穆感。

The Hart College construction started in 1915 and was completed in 1920, which was donated by Canada Methodist Church in memory of Dr. Virgil Chittenden Hart, the earliest American to preach in southwest China. It was originally the College of Science Building of West China Union University, and it is now the Chengdu American Overseas Study Center. The building is the combination of Chinese and Western architectural styles. It has a square tower with three eaves and four corners and the single eaves saddle roof. It is creative to build Western towers on a traditional Chinese roof. With multiple eaves the Hart College building is the building that has the most eaves in Huaxiba. The square tower on the gatehouse adds solemnity to the whole building.

20世纪50年代初，合德堂被命名为第四教学楼。

In the early 1950s, the Hart College was named the Fourth Teaching Building.

2001年11月，第四教学楼被成都市政府批准列为成都市首

上图：高高耸立的合德堂塔楼在华西坝建筑群里显得特别引人注目，其左边的建筑是嘉德堂。图片来源：四川大学档案馆。
The tall Hart College tower is particularly striking in the Huaxiba building complex, and the building on its left is the Biology Building. Photo source: Sichuan University Archives.

右图：如今的合德堂依然是华西坝上最具特色的老建筑。戚亚男摄于2012年。
Now, the Hart College is still the most distinctive old building in Huaxiba. Photo by Qi Yanan in 2012.

批文物建筑之一。

2002年12月，第四教学楼被四川省政府批准列为四川省第六批文物保护单位。

2013年5月，学校恢复命名第四教学楼为合德堂；同年5月，该楼被国务院确定为全国重点文物保护单位。

In November, 2001, the Fourth Teaching Building was confirmed as one of the first Cultural Relic Buildings in Chengdu by the Chengdu Municipal Government.

In December, 2002, the Fourth Teaching Building was confirmed as the sixth batch of Cultural Relic Protection Units in Sichuan Province by the Sichuan Provincial Government.

In May, 2013, the Hart College building was confirmed as a National Key Cultural Relic Protection Unit by the State Council.

引西医入川的美国传教士赫斐秋
Virgil Chittenden Hart, Introduced Western Medicine into Sichuan

此楼是为了纪念引西医入川的加拿大传教士赫斐秋，故命名为赫斐楼，也就是合德堂。

The Hart College building was built to memorize Dr. V. C. Hart, who was a Canada Methodist and introduced Western Medicine into Sichuan.

赫斐秋1840年出生在美国纽约，1865年，被派遣到中国传教。赫斐秋被派往芜湖、南京、镇江、重庆等地传教。1888年，赫斐秋回到美国后，出版了《中国西

部——佛教圣地峨眉山游记》一书，第一次向西方介绍四川的峨眉山。此时因他的身体状况不好，遂前往加拿大安大略省休养。经过两年多的休养，赫斐秋的身体渐渐康复起来了，尽管他50来岁了，但仍然渴望继续从事海外传教工作。加拿大美道会邀请赫斐秋加入该会，负责海外医学传教。赫斐秋凭借他在中国20多年的传教经历，为教会选择了到中国西部中心——成都进行医学传教。教会同意了他的方案，并派遣了启尔德、斯蒂文森两位医学传教士以及另外一位传教士组成传教团，由赫斐秋负责带领他们到四川进行医学传教。

V. C. Hart was born in New York in 1840. He was sent to China to preach in 1865. He went to Wuhu, Nanjing, Zhenjiang and Chongqing to preach. In 1888, after Dr. Hart returned to the United States, he published the book *Travel Notes to Mount Emei-A Buddhist Holy Land in Western China*, which first introduced Mount Emei in Sichuan to the West. At this time, his health situation was not good, and he went to Ontario, Canada to recuperate. After more than two years of recuperation, Dr. Hart gradually recovered. Although he was in his 50s, he was still eager to continue in overseas missionary work. The Canadian Methodist Church invited him to join the society, in charge

1891年赫斐秋带领加拿大卫理传教会第一批来四川的传教士到成都进行医学传教。前排从左起：赫斐秋夫妇、斯蒂文森。后排从左起：何忠义夫妇、启尔德夫妇。图片来源：《我们的中国西部传教团》。

In 1891, Dr. Hart led the first group of Canadian Methodist missionaries to Sichuan to carry out medical missionary work in Chengdu. Front row, from left: Dr. V. C. Hart, Mrs. Hart, Dr. D. W. Stevenson. Back row, from left: Mr. G. E. Hartwell and Mrs. Hartwell, Dr. O. L. Kilborn and Mrs. Kilborn. Photo source: *Our West China Mission*.

of overseas medical missions. With his more than 20 years of missionary in China, Dr. Hart chose to conduct a medical mission in Chengdu, the center of western China. The church agreed to his plan and sent Dr. O. L. Kilborn and Dr. D. W. Stevenson, two medical missionaries and another missionary to form a mission, headed by him and led them to Sichuan for medical missions.

1891年10月4日，赫斐秋带领传教团包括家属一行人离开加拿大，于11月3日抵达上海。由于时局动荡，赫斐秋决定传教团暂时在上海停留一段时间，利用这段时间学习中文和讨论怎样进行医学传教工作。传教团在上海待了三个多月后，次年2月16日，他们乘船逆长江而上，进入四川，然后经重庆历经3个多月的行程终于于5月21日到达了成都。

On October 4, 1891, Dr. Hart led the mission away from Canada, the group including their family members, arrived in Shanghai on November 3. Because of the turbulent situation, he decided that the mission would stay in Shanghai for a period of time to learn Chinese and how to conduct medical missions. After studying in Shanghai for three months, on February 16 of the following year, they sailed up the Yangtze River into Sichuan, passing through Chongqing, and reached Chengdu on May 21 after more than three months of travel.

刚到成都就遇到当地霍乱的暴发，启尔德的夫人因感染霍乱而去世了。赫斐秋只好带领他们暂时撤离成都，到几十公里以外的郊区山上躲避瘟疫。一个多月以后，疫情慢慢退去，赫斐秋又带领同行人员回到城里。他们购置地产，修建了学校、教堂和诊所。

Just arrived in Chengdu, Dr. Hart and his team encountered the outbreak of cholera, and Dr. Kilborn's wife was infected with cholera and died. Dr. Hart had to lead them to temporarily leave Chengdu, to dozens of kilometers away to escape the plague. More than a month later, when the epidemic slowly receded, Dr. Hart led his team back to the city. They bought real estate and built schools, churches and clinics to carry out missionary work.

在中国工作了几十年的赫斐秋深知中国人喜欢读书，他也清楚书籍在传播文化方面的重要性，而书籍的传播离不开印刷。当时在四川还没有现代印刷技术，书籍都是使用传统的手工工艺印制，效率不高，不能印制英文、藏文，更不能印制彩色的印刷品。

Having worked in China for decades, Dr. Hart knew that the Chinese love to read. He understood the importance of books in spreading culture, and books can't do without the printing press. At that time, there was no modern printing technology in Sichuan. Books were printed by traditional manual process, which was inefficient and could not be printed in English, Tibetan, or even color.

1897年，赫斐秋利用回美国休假的机会募集到一笔经费并用这笔钱买了两台印刷机，亲自押运回四川，在乐山开设了全川第一家具有近代印刷技术的印刷馆。该馆1904年迁往成都，改为华英书局，1951年更名为成都印刷厂。

In 1897, Dr. Hart took the opportunity of returning to the United States on vacation to raise money to buy two printing presses, escorted them back to Sichuan, and opened the first modern printing technology printing plant of Sichuan in Leshan. The plant was moved to Chengdu in 1904 and was named Canada Methodist Mission Press. In 1951, it was renamed Chengdu Printing Factory.

赫斐秋、启尔德等人在1892年开办的诊所可谓是成都第一家西医诊所，该诊所后来发展成为仁济医院（男医院，女医院）。经过几十年的岁月，仁济医院分成了两个医院。一个是由仁济医院分迁到华西协合大学的教学医院，就是后来的四川医学院附属医院、华西医科大学附属医院，现在的四川大学华西医院（简称华西医院），目前华西医院是西南地区最大的医院。另一个就是留在仁济医院原址的现在的成都市第二人民医院。

The clinic opened by Dr. Hart, Dr. O. L. Kilborn and others in 1892, was the first western medicine clinic in Chengdu, which later developed into the C. M. M. Hospital (for men, for women and children). After decades, the Hospital was divided into two hospitals. One was transferred to the teaching Hospital of West China Union University, later the Affiliated Hospital of Sichuan Medical College, the Affiliated Hospital of West China Medical and Science University, and now the West China Hospital of Sichuan University. The West China Hospital is currently the largest hospital in southwest China. The other one is the present Chengdu Second People's Hospital, which stayed on the original site of the C. M. M. Hospital.

西南地区科学先驱戴谦和博士
Daniel Sheets Dye, a Scientific Pioneer in Southwest China

合德堂是华西协合大学理学院最早的教学大楼，该楼是中国西部最早培养理科人才的地方。最早来这里从事理科教学和研究工作的是美国人戴谦和，他是创办华西协合大学的元老之一，也是在华西坝上工作时间最久的一位外教，从1908年来中国，直到1949年才离开，他曾五次出任理学院院长，数次担任数理系主任。

The Hart College building is the earliest teaching building of the College of Science, and it is the earliest training place for science talents in western China. The American Daniel Sheets Dye is the first to be engaged in science teaching and research work at here. Dr. Dye is one of the founders of West China Union University, also one of the longest-serving foreign teachers in West China Union University. He came to China in 1908, did not leave until 1949 and served as Dean of the College of Science five times, and as dean of mathematics department several times.

1884年2月7日，戴谦和生于美国俄亥俄州，从美国康奈尔大学毕业以后，他在美国威斯康辛大学教了一年书，后远渡重洋于1908年10月26日来到了中国。他在上海沪江大学教了一年书后，便来到成都，参与华西协合大学的创办。1910年大学开办时，戴谦和就担任物理学教授。

D.S. Dye was born on February 7, 1884 in Ohio, USA. After graduating from Cornell University, he taught at the University of Wisconsin for one year, and then came to China on October 26, 1908. After teaching in the University of Shanghai for one year, he came to Chengdu to participate in the establishment of West China Union University. When the university opened in 1910, he was a professor of physics.

戴谦和待人如同他的中文名字一样非常谦和，与任何人交谈他都总是笑嘻嘻的。他的学生祝绍琪曾回忆道："我们读他的普通物理，他授课时那副卖劲的神情，至今还深深地印在我的脑海里，他的中国话讲得不十分好，而上课他偏喜欢讲中文，有时他的话表达不出他的意思，那副着急的样子，常使我们偷偷发笑，当他讲授到分子运动的时候，他闭着眼踮着脚，偏着头，右手在空中晃动，口里说着：'拍，拍，拍……那个分子的跳动'，这幅动人的镜头，我们至今还能学着做这些动作。"

Dr. Dye is as mild-mannered as his Chinese name. He always smiles when he talks to anyone. His students Mr. Zhu Shaoqi once recalled, "We read his general physics, and the eager look he used to give lectures still remains in my mind. He didn't speak Chinese very well, but in class he preferred to speak Chinese. Sometimes his words didn't express what he meant, so he looked very worried and anxious, that often made us snicker. When he taught molecular motion, he closed his eyes, tiptoed, tilted his head, and his right hand moved in the air. He said, 'Pa. Pa. Pa…The beating of the molecule', this moving scene, we can still learn to do these actions now."

为人谦和的戴谦和做事很认真、严谨。当年他刚来华西坝时，负责校园的建设工作，在工人修筑道路和栽培树木时他发现了不少问题，由此开始研究成都的地质、土壤、气候、风向、雨量、地下水以及树木的生长状况。然而那时既无资料可查询，又无人可问，他只有从实践中去寻求答案。经过12年的潜心研究，他不但对如何植树有了经验，而且对四川地层的构造、成都盆地的成因也提出了自己的观点。鉴于他对成都地质学的贡献，1943年英国皇家地质学会聘请他为该会会员。

The modest Dr. Dye was very serious and rigorous. When he first came to Huaxiba he was responsible for the construction of the campus, especially the construction of roads and the planting of trees. He found many problems, so he began studying Chengdu's geology, soil, climate, wind direction, rainfall, groundwater, and the physiology and growth of trees. But then, with no data to consult and no one to ask, he had to look for answers in practice. After 12 years of painstaking research, he not only had experience on how to plant trees, but also put forward his own view on the formation of Sichuan strata and the origin of Chengdu Basin. In view of his contribution to the geology of Chengdu, in 1943 he was offered membership of the Royal Geological Society.

戴谦和在华西坝上以博学多才而闻名，他是一位物理学教授，却对地理、地质、气象、植物有研究，特别是在考古学方面有杰出的成果。1914年，校园新安装了一个机械大钟，旧的敲钟显然就没有用处了，但戴谦和独具慧眼，立刻建议学校

成立博物馆，收集有历史价值的物品，那口旧钟就成了博物馆的第一件藏品。随后，戴谦和与同事陶然士、叶长青开始有目的地收集古物。合德堂建成以后，这些文物都陈列在里面，到1931年已收集有6 000余件，戴谦和为博物馆的负责人。1932年，美国人类学家葛维汉来到华西坝后，学校正式成立了古物博物馆，陈列在合德堂的文物全部搬迁入懋德堂二楼，葛维汉出任馆长。该博物馆就是现在的四川大学博物馆，是目前中国高等院校中唯一的一所综合性博物馆。

戴谦和（右）与同事在合德堂实验室里做物理学实验。图片来源：四川大学博物馆。

Dr. Dye (right) and his colleague are doing physics experiment in laboratory of the Hart College. Photo source: Sichuan University Museum.

Dr. Dye is known for his erudition in Huaxiba. He was a professor of physics, but he also made research on geography, geology, meteorology, plants, especially outstanding achievements in archaeology. In 1914, a new mechanical clock was installed in the university, and the old bell was obviously useless. But he had unique insight. He immediately suggested that the university set up a museum to collect items of historical value. The old bell became the first collection of museum. Subsequently, Dr. Dye and his colleagues, Mr. Thomas Torrance and Mr. Huston Edjar, began to collect antiquities on purpose. When the Hart College was built, the artifacts were all on display in it, and by 1931, more than 6,000 pieces were collected. Dr. Dye was the head of the museum. In 1932, after the American anthropologist Dr. David Crockett Graham came to Huaxiba, the university officially established the Antiquities Museum. All the artifacts displayed in the Hart College were moved to the second floor of the Library and the Museum Building, with Dr. Graham as the curator. The museum is now the Sichuan University Museum, which is the only comprehensive museum among institutions and universities in China.

尽管戴谦和的中文说得不够流利,但他的夫人简恩·鲍德敦的中国话却讲得很好,也能书写中文,她曾用中文编写过一本数学教科书。鲍德敦1886年4月21日生于美国马里兰州,她是戴谦和的大学同学,1916年来四川当数学老师。1919年5月19日戴谦和与她在成都结婚。鲍德敦的爱好与丈夫不同,无论晴天或是雨天,一早一晚,她总是带着一个望远镜在校园里观鸟。1926年,她还撰写了《华西协合大学校园里20种常见的鸟类》在专业杂志上发表。

Although Dr. Dye was not fluent enough in Chinese, his wife, Ms. Jane Canby Balderston, who wrote a math textbook in Chinese, speaks and writes Chinese well. She was born in Maryland USA on April 21,1886. She was a classmate of Dr. Dye at the university. She came to Sichuan in 1916 to teach mathematics. Dr. Dye married her in Chengdu on May 19, 1919. Mrs. J. B. Day's hobby was different from her husband. Whether it was sunny or rainy, in morning and night, she carried a telescope with her to watch birds on campus. In 1926, she wrote an article entitled *Twenty Common Birds of the West China Union University Campus*, which was published in a professional magazine.

从合德堂走出来的天文学家李珩
Famous Astronomer Li Heng Coming Out of Hart College

在理学院早期毕业的几位理科学生中有一位特别出众的学生叫李珩,他是我国现代天文事业的奠基人之一。李珩字晓舫,1898年出生于成都,父亲经商。李珩小时候,每当夏天的晚上母亲带着他坐在院子里乘凉时,他仰望夜空,漫天闪烁的星星以及偶尔划过天空的流星都让他好奇,而母亲给他讲的牛郎织女的故事更引起了他的遐想。1910年,每隔76年就回归的哈雷彗星再次出现,12岁的李珩目睹了这奇特的天文现象,此时一个梦想在他小小的心里扎下了根——那就是将来要从事天文学的研究。

One of the several outstanding science students who early graduated from the College of Science was Dr. Li Heng, who was one of the founders of modern astronomy in China. Dr. Li Heng's style name is Li Xiaofang. He was born in Chengdu in 1898. His father was in business. Every summer night when he and his mother sat in the yard to enjoy the cool, he looked up at the night sky. The star flashing over the sky and the occasional meteor across the sky made him curious, and the story of the Cowherd and the Weaver Girl his mother told him aroused his reverie. In 1910, Halley's Comet reappeared which returned every 76 years, and 12-year-old Li Heng

天文学家李珩。图片来源：1936年《山大年刊》。
Famous astronomer Li Heng. Photo source: *Shandong University Annual 1936*.

witnessed this strange astronomical phenomena. This time, a dream took root in his little mind, that is to study astronomy in the future.

1917年李珩高中毕业，他想实现学天文学的梦想，然而当时现代科学刚刚传入中国，没有专门的学校可以学天文，但数学是天文学最重要的基础知识，因此李珩就考入华西协合大学理科学数学，为日后学天文学做准备。1922年李珩毕业后留校任教。李珩因执意要从事天文学的事业，他父亲想要他子承父业的愿望便落空了，因此他父亲不再为他继续学习提供经济支持。不久他到重庆第二女子师范学校教数学和英语，其目的是积攒出国学天文学的费用。

When Li Heng graduated from high school in 1917, he wanted to realize his dream of learning astronomy. However, modern science had just been spread into China and there was no special college to learn astronomy, but mathematics was the most important basic knowledge of astronomy. Therefore, He was admitted to West China Union University to study Mathematics in the College of Science to prepare for learning astronomy in the future. In 1922, Li Heng graduated and stayed at college to teach. He insisted on engaging in astronomy career, but his father wanted him to follow his father's career, but failed. So his father had no longer provided economic support for his further study. Soon he went to Chongqing Second Women's Normal School to teach mathematics and English, to accumulate money to go abroad to study astronomy.

1925年，李珩来到法国巴黎大学学习。他与在巴黎大学留学理科的几位同学了解到，自从20多年前中国学生来巴黎大学求学以来，到现在没有一个组织把留学的同学聚集在一起相互交流学习和生活。李珩和同学就把40多位在巴黎大学留学理科的中国学生召集在一起，于1926年4月16日成立了巴黎大学理科中国同学会，每个月同学会都要开一次演讲会，由会员轮流主讲，或请巴黎大学的教授来演讲，当年李珩演讲的题目有"谈天"和"近世宇宙观"等。

In 1925, Li Heng came to France to study at the University of Paris. He and several of Chinese science students at the University of Paris felt that since Chinese students came to the University of Paris more than 20 years ago, there had not been an organization to bring Chinese students together to communicate their study and life experience. They gathered more than 40 Chinese students studying in the University of Paris Science Department, and on April 16, 1926, established Chinese Alumni Association of Sciences of the University of Paris. The association held a lecture once a month. The member of association took turns or ask the professor of the University of Paris to make a speech. In those days, Li Heng lectured the topics of "Talk about Sky" and "Recent View of the Universe" etc.

1928年，李珩和同学创办了《巴黎大学理科中国同学会杂志》，该杂志专门刊登同学会成员撰写的学术文章，李珩担任该杂志的编辑，他自己在杂志上发表了《N度空间之一瞥——超越球》《光之分析与其在天文学上之应用》等有关天文学的文章。

In 1928, Li Heng and schoolmates founded *Journal of Chinese Alumni Association of Sciences of the University of Paris*. The magazine published specifically academic articles written by the association members. Li Heng served editor of the magazine. He published in the magazine *A Glimpse of the N Degree Space—Hypersphere*, *Light Analysis and Its Application in Astronomy* and other articles about astronomy.

两年后，李珩在巴黎大学的校园里巧遇了他在重庆第二女子师范学校任教时教过的一位四川岳池县女学生——罗玉君，她在法国巴黎大学学文学，不久两人相爱了。一天，李珩对罗玉君说，自己因为酷爱天文学，违背了他父亲想要他经商的愿望，家里很早就断绝了提供自己读书的经济来源，而自己在国内教书积攒的钱也快用完了，只好回国谋生。罗玉君对李珩说：“你的学习成绩那么好，考公费留学应该不成问题。”后来李珩果然考取了公费留学，到法国里昂大学学习天文学。李珩离开巴黎到了里昂后给罗玉君写了一封很浪漫的信，信中这样写道："从现在起，我每天给你写一封信，等我写满一千封信，就向你求婚。"在这期间，罗玉君除了去巴黎大学上课外，已经开始翻译法国著名的文学作品。等李珩获得了博士学位从里昂回到巴黎时，罗玉君翻译的法国作家都德的文学作品《婀丽女郎》已经在国内的商务印书馆出版发行了。

Two years later, Li Heng met a female student in the campus of the University of Paris. She is Ms. Luo Yujun, from Yuechi County, Sichuan Province. She studied literature at the University of Paris. When Li Heng was teaching at Chongqing Second Women's Normal School, she was a student of him. Soon they fell in love. One day, he told Ms. Luo Yujun that because he loved astronomy and violated his father's wishes that he should go into business, his family had long before cut off his source of economics for education, and the money accumulated from teaching in China was almost running out. So he had to return to China to make a living. Smart Luo Yujun told him, "your academic performance is so good that it is not a problem to study abroad at public expense". Sure enough, Li Heng got the public allowance to study astronomy at the University of Lyon in France. After Li Heng left Paris to Lyon, he wrote a very romantic letter to Ms. Luo Yujun, saying, "From now on, I will write you a letter every day. When I write a thousand letters, I will propose to you." During this period, in addition to study at the University of Paris, Ms. Luo Yujun had began to translate famous French literary works. By the time Li Heng received his doctorate and returned to Paris from Lyon, Ms. Luo Yujun's translation of French writer Tudor's Literary work *Queen Araya* has been published and distributed by the commercial press in China.

1932年李珩与罗玉君在巴黎结婚了，婚后生了一个女儿。1933年10月，李珩一家三口回国到了山东青岛，夫妻俩都在山东大学教书，李珩还被青岛观象台聘为特约研究员，罗玉君在教学之余继续从事她钟爱的翻译工作，翻译法国作家司汤达的小说《红与黑》。

Dr. Li Heng married Ms. Luo Yujun in Paris, where they gave birth to a daughter. In October, 1933, Dr. Li Heng and his family returned to Qingdao, Shandong Province, both teaching at Shandong University. Dr. Li Heng was also hired as a special researcher by Qingdao Observatory. In addition to teaching, Ms. Luo Yujun continued to do her favorite translation of *The Red and the Black* by French writer Stendhal.

抗战期间，李珩、罗玉君夫妇带着孩子回到了故乡四川，夫妇二人分别在成都华西协合大学理学院与文学院任教。尽管成都没有天文台供李珩深入研究天文学，但他还是不放弃对天文的热爱，他把重点放在了天文科普工作上。从1937年到1949年，他在国内的10多种杂志上发表了30多篇有关天文学方面的科普文章，如《宇宙线》《日食之计算与中国未来之日食》《银河系测度之演进》《星云演进中之危险时期》《太阳亦如土地星之有光环乎？》《天文家之工具》《日食漫谈》《星际物质》《太阳之物理的研究》《卅年来天文学之重要进步》《天体物理学之进展》《战争期间天文学进步之一瞥》《天文学研究对于哲学及人生观之影响》等等。

During the Counter-Japanese War, Dr. Li Heng and his wife returned to their hometown of Sichuan with their children. The couple respectively taught at the College of Science and the College of Literature of West China Union University in Chengdu. Although there was no astronomical observatory in Chengdu for Dr. Li Heng to study astronomy deeply, he still did not give up his love for astronomy. He focused on astronomy science popularization. From 1937 to 1949, he published more than 30 popular science articles on astronomy in more than ten journals of domestic, such as *Cosmic line*, *The Calculation of the Eclipse and Chinese Future Eclipse*, *The Evolution of the Galaxy Measure*, *The Dangerous Period of Nebula Evolution*, *Has the Sun a Halo Like Saturn Star?*, *The Tool of an Astronomer*, *Talk about Solar Eclipse*, *Interstellar Matter*, *The Study of the Sun Physics*, *The Important Progress in Astronomy for 30 years*, *The Progress of Astrophysics*, *A Glimpse of Astronomy Progress During the War*, *The Impact of Astronomy Research on Philosophy and Outlook on Life* and so on.

李珩在《天文学研究对于哲学及人生观之影响》文章里从哥白尼的日心说、牛顿的万有引力和爱因斯坦的相对论谈起，向读者介绍了天文学的研究对于人类历史的影响与启示。他写道："天文学对于其他学科之贡献，质量是二方，均可以称道。""天文学在科学演进史中，实居重要之地位，由天文学之研究，始知自然有规律可循。""天文学更昭示现象虽复杂，而定理却甚简单。""天文学既示吾人以如是伟大而奇妙之宇宙，则吾人本此宇宙观以对世事，眼界胸襟当必不同寻常，是又天文学之影响于哲学者也。"

In his article *The Impact of Astronomy Research on Philosophy and Outlook on Life*, Dr. Li Heng introduced the influence and enlightenment of astronomy research on human history by talked about Copernicus's heliocentric theory, Newton's universal gravitation and Einstein's theory of relativity. He wrote, "The quality of astronomy's contribution to other disciplines is from two sides, which can be praised." "Astronomy plays an important role in the evolutionary history of scientific evolution. From the study of astronomy, we first know that nature has laws to follow." "Astronomy shows that although the phenomenon is complex, the theorem is very simple." "Astronomy not only shows that our human being have such a great and wonderful universe, but also our own view of the universe to the world, and our vision must be unusual. It is the influence of astronomy also on philosophies."

李珩在忙于自己的天文事业时也没有忘了帮助夫人完成她翻译《红与黑》的心愿。当年罗玉君在法国留学期间就开始翻译这部有四五十万字的小说，因为当年她还在留学，小说也就只翻译了五六章。留学之后，罗玉君带着这部没有翻译完成的手稿回国到青岛继续翻译。没想到抗日战争爆发了，罗玉君被迫中断了翻译，随丈夫与孩子流亡到四川。然而，罗玉君的翻译工作因为生计的忙碌不能持续地进行，只能在暑期抽时间翻译。罗玉君曾这样说过："我得感谢陈翔鹤先生，他送给我一本英文《红与黑》译本，使晓舫得据此仔细校对了一遍。自然，我更应该感谢晓舫，他有时给我鼓励，有时给我讥讽，他说我在刺绣《红与黑》，在解剖《红与黑》，永远不能完成《红与黑》。"1947年，罗玉君终于翻译完成了《红与黑》这本世界名著，随后不久此书便正式出版了。

Dr. Li Heng was busy with his astronomical career and did not forget to help his wife to complete her wish of translation *The Red and the Black*. When Ms. Luo Yujun was studying in France, she began to translate this novel of 400,000 or 500,000 words. Because she was busy studying, she only translated five or six chapters of the novel. After studying abroad, she returned to Qingdao with the unfinished manuscript to continue the translation. Unexpectedly, the Counter-Japanese War broke out, and Ms. Luo Yujun was forced to interrupt the translation and fled to Sichuan with her husband and children. However, her translation work could not be carried out continuously because of the busy livelihood, and she could only take time in the summer vacation to translate. She once said, "I'd like to thank Mr. Chen Xianghe. He gave me an English translation of *The Red and the Black*, so that Xiaofang could refer to it to proofread my translation carefully. Naturally, I should thank Xiaofang. He sometimes gave me encouragement, sometimes gave me sarcasm. He said that I was embroidering *The Red and the Black*, dissecting *The Red and the Black*, could never complete *The Red and the Black*". In 1947, Ms. Luo Yujun finally finished the translation of the world famous book *The Red and the Black*, which was officially published shortly.

1948年9月，李珩应邀到美国普林斯顿大学讲学，在讲学的近一年时间里，他看到一些在美国稍有成就的中国人已经计划留在美国，而他却说："吾辈受国家培植甚厚，岂能在患难间各自作自身安适打算？但看战争期间欧洲各国科学家逃此邦获得优越位置者，十、九现已归国受苦，吾辈岂能无动于衷！"第二年7月李珩如期回国。

In September, 1948, Dr. Li Heng was invited to give lectures at Princeton University in the United States. During the nearly year-long lecturing, he saw that some Chinese with a little achievements in the United States planned to stay in the United States. He said, "My generation is very well cultivated by our country, so in the national disaster period, how could we make our own comfort plans? But when scientists from European countries fled to U.S.A. and obtained superior position during the war, most of them have returned to their country and suffered. How can we be completely indifferent!" Dr. Li Heng returned to China in July of the next year as scheduled.

1951年,李珩应中国科学院院长郭沫若的邀请,出任中国科学院紫金山天文台研究员,并先后任上海佘山观象台和徐家汇观象台负责人。1962年,中国科学院决定成立上海天文台,李珩被任命为上海天文台第一任台长一直到1981年。

In 1951, Dr. Li Heng was invited by Guo Moruo, President of the Chinese Academy of Sciences and became a researcher at the Purple Mountain Observatory of the Chinese Academy of Sciences, and successively the head of Shanghai Sheshan Observatory and Xujiahui Observatory. In 1962, the Chinese Academy of Sciences decided to establish the Shanghai Observatory, and Dr. Li Heng was appointed the first director of the Shanghai Observatory until 1981.

华西坝老建筑的前世今生
The Stories of the Historic Architecture of Huaxiba

拾贰

教育学院

育德堂

The Cadbury School of
Education Building

育德堂又名教育学院大楼，或嘉弟伯教育学院，1928年竣工，由英国嘉弟伯氏捐资修建。从当年拍摄的照片里可以明显地看出该楼西面少建了侧翼。直到1948年，西康省主席刘文辉出资修建了西面的侧翼，使整栋大楼看上去才完整。该建筑的特点是大楼的门廊是由西式半圆形拱券门构成的，台阶为中式垂带踏跺式。一、二层的屋檐的斗拱是从墙体上直接挑出的简化一斗三升结构。一直以来，人们都不知道为什么当年修建育德堂时没有建西面侧翼，有人猜测是因为经费不足。

　　The Cadbury School of Education Building or the College of Education Building, donated by the British Mr. George Cadbury was completed in 1928. From the photos taken in that year, it is obvious that the west wing of the building was missing. Until 1948, Liu Wenhui, the then Chairman of Xikang Province, donated money to build the west wing, making the whole building look complete. The building feature is the porch, which is made of western style semicircular arches with Chinese vertical belt treading steps. The bucket arches on the eaves of the first and second floors are a simplified a bucket of three liters of brackets directly from the wall. It has long been unknown why the west wing was not built, and speculation was underfunded.

　　2018年，英国人赵安祝来四川大学捐赠他高祖父荣杜易当年设计的华西协合大学建筑图纸，他也拷贝了一份给笔者。从育德堂的设计图纸上笔者终于明白了当年为什么没有修建西面

1928年落成的育德堂西翼没有建成，是因为那块土地没有被购买到。图片来源：耶鲁大学。
The Cadbury School of Education Building was completed in 1928. But the west wing was not built because the land had not purchased. Photo source: Yale University.

如今育德堂为四川大学华西校区基础医学院的教学楼之一。咸亚男摄于2005年。
Now the Cadbury School of Education Building is one of the teaching buildings of West China School of Basic Medical Sciences & Forensic Medicine, Sichuan University.
Photo by Qi Yanan in 2005.

侧翼。在设计图纸上，西头部分画了一条虚线，旁边写有"没有购买到的坟地"，原来育德堂的西面是设计在还未购买到坟地使用权的地方，因而当初建该楼时只好放弃了修建西面侧翼。

In 2018, a British friend Mr. Andrew George came to Sichuan University to donate the architectural drawings of West China Union University designed by his great-great-grandfather Rowntree. He also provided a copy of the architectural drawings to the author. From the design drawings of the Cadbury School of Education Building, the author finally understood why in those years the west wing was not built. On the design drawings, the west head section drew a dotted line next to the words "graves not purchased". It turned out that the west wing of the building was designed on the graves not yet purchased, so the west wing had to be given up when the building was built.

20世纪50年代，该楼被命名为第五教学楼，现在设有四川大学华西校区的基础医学院教研室。

In the 1950s, the building was named the Fifth Teaching Building, and now it is the building of West China School of Basic Medical Sciences & Forensic Medicine, Sichuan University.

2001年11月，第五教学楼被成都市政府批准列为成都市首批文物建筑之一。

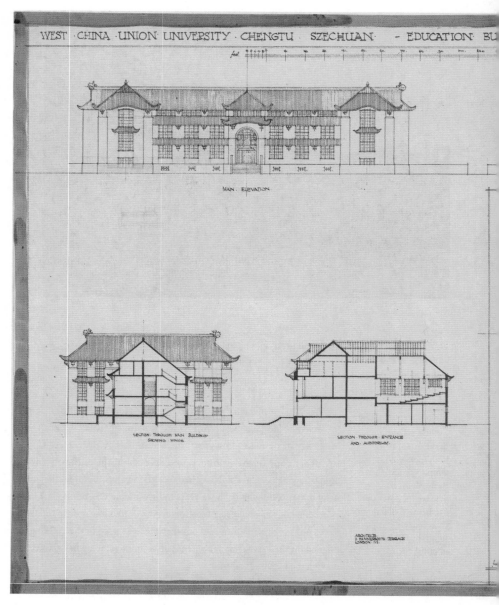

荣杜易设计的育德堂正立面图和剖面图,以及育德堂在校园里的具体位置,还有虚线标出的育德堂西翼部分正好在未购买到坟地使用权的地方。赵安祝提供。

The front elevation and section drawings of the Cadbury School of Education Building designed by Mr. Rowntree, as well as the specific location of the building in the campus. The west wing of the building marked by the dotted line was located on graves not purchased. Andrew George provided.

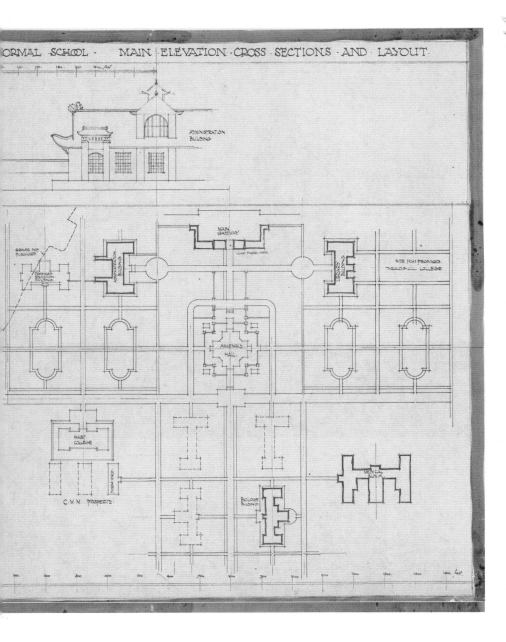

荣杜易设计的一款育德堂手稿。赵安祝提供。
This is sketch of the Cadbury School of Education Building designed by Mr. Rowntree. Andrew George provided.

荣杜易设计的这款育德堂手稿最终被采用。赵安祝提供。
At last, this sketch of the Cadbury School of Education Building designed by Mr. Rowntree was adopted. Andrew George provided.

2002年12月，第五教学楼被四川省政府批准列为四川省第六批文物保护单位。

2013年5月，学校命名第五教学楼为育德堂；同年5月，该楼被国务院确定为全国重点文物保护单位。

In November, 2001, the Fifth Teaching Building was confirmed as one of the First Cultural Relic Buildings in Chengdu by the Chengdu Municipal Government.

In December, 2002, the Fifth Teaching Building was confirmed as the sixth batch of Cultural Relic Protection Units in Sichuan Province by the Sichuan Provincial Government.

In May, 2013, the Fifth Teaching Building was named the Cadbury School of Education Building by Sichuan University, and was confirmed as a National Key Cultural Relic Protection Unit by the State Council.

中国乡村教育的先行者傅葆琛
Fu Baochen, a Pioneer of Rural Education in China

1910年华西协合大学开办时就设有教育科，主要为西部地区的教会小学和中学培养师资力量。但是到了1933年，华西协合大学向中国政府教育部立案成功后，按照国民政府教育部的规定，私立学校不得开办师范教育，大学在1932年就撤销了教育科，同时在文学院里增加了教育系。

Since West China Union University opened in 1910, it established an Education School, mainly training teachers for church primary and secondary schools in the western region. However, in 1933, after West China Union University successfully registered with the Ministry of Education of the Chinese government, according to the regulations of the Ministry of Education of the National Government, private schools were not allowed to open normal education. The university revoked the Education School in 1932 and added the Education Department in the School of Arts.

抗战期间，国民政府教育部令私立大学不得办教育系，该系随后改名为乡村建设系，由乡村教育专家傅葆琛主持。

During the Counter-Japanese War, the Ministry of Education of the National Government banned private universities from running the department of education. The Education Department in the School of Arts was later renamed the Department of Rural Construction, presided over by Fu Baochen, a rural education expert.

傅葆琛1893年出生于四川华阳县永安乡，4岁时就随父母离开家乡到北方生活。1916年，傅葆琛从北平清华大学留美预备学校毕业后，到美国俄勒冈州农科大学森林学院学习。他毕业时，正值第一次世界大战。他原本打算到美国东部继续学习后，再到新英格兰各地考察当地的农业。傅葆琛在纽约遇到了国际青年会招聘干事准备到法国，去为那里从事战争后方工作的10多万从中国招募去的工人服务。出于为同胞帮忙，以及能借此机会"看看欧战和欧洲农民的生活"的目的，1918年秋，傅葆琛应邀去了法国，与先期应聘的晏阳初等人从事华工的教育工作。

Fu Baochen was born in 1893 in Yong'an Township, Huayang County, Sichuan Province. At the age of 4, he left his hometown to live in the north with his parents. In 1916, after graduating from Tsinghua University Preparatory School for Studying in the United States, Fu Baochen went to study at the Forest College of Agricultural University of Oregon. When he graduated during the World War I, he had planned to observe and study agriculture in New England after continuing his studies in the Eastern United States. In New York, the All Nations Youth Association was recruiting officers to go to France to serve more than 100,000 workers recruited from China to work in the war rear. To provide help for compatriots and to get the opportunity to "see the European war and the life of European farmers", in the autumn of 1918, Fu Baochen was invited to France to engage in the education of Chinese workers with Y. C. James Yen, and others who were previously invited.

上图：乡村教育家傅葆琛。图片来源：《农业周报》1931年第1卷第11期。
Rural education educator Fu Baochen. Photo source: *Agricultural Weekly*, Volume 1, Issue 11, 1931.

1921年，傅葆琛结束在法国的华工教育工作回到美国，在耶鲁大学森林学院继续深造。然而，他在法国的经历却让他改变了对农业的爱好，他对朋友说："我没有到法国办华工教育之前，只晓得我自己的教育要紧。既到法国之后，才晓得别人的教育比我的教育更要紧。"随后，傅葆琛放弃了林业学的研究，转入康奈尔大学农业研究院研究乡村教育，并于1924年获得"乡村教育博士"学位。

In 1921, Fu Baochen returned to the United States after finishing his education work of Chinese laborer in France and continued his further study at the Yale University College of Forestry. However, his experience in France made him change his interest in agriculture. He said to his friends, "Before I went to run education work of Chinese laborer in France, I only knew that my own education was important. After I arrived in France, I realized that other people's education was more important than mine." Fu Baochen then abandoned his research in forestry and transferred to the Agriculture Research Institute of Cornell University to study Rural Education, and received the degree of "Doctor of Rural Education" in 1924.

而此时在国内，当年在法国与傅葆琛共事的晏阳初早已从美国留学归来，与朱其慧、陶行知于1923年在北平成立了中华平民教育促进会，晏阳初任总干事。得知傅葆琛毕业，晏阳初便邀请傅葆琛回国任该会乡村教育部主任。

At this time in China, Dr. Y. C. James Yen, who worked together with Fu Baochen in France, had already returned from studying in the United States. He, together with Ms. Zhu Qihui and Mr. Tao Xingzhi, founded the Chinese Civilian Education Promotion Association in Beiping in 1923. Dr. Y. C. James Yen served as the director general. Knowing that Dr. Fu Baochen graduated, he invited Dr. Fu Baochen to return home to serve as director of the director of Rural Education of the association.

1924年，傅葆琛到任后，立刻开展工作。他为宣传推广乡村平民教育作《为什么要办乡村平民教育？》的讲演，他说："中国的前途，还要靠这大多数乡村的人民。他们强，中国也就强；他们富，中国也就富；他们弱，中国也就弱；他们穷，中国也就穷。""我们要想改造中国，第一步应该做的事，就是要提高民智，普及教育。"接着傅葆琛花费了近一年的时间编辑出版《农民识字课本》，出版《农民周刊》。

In 1924, Dr. Fu Baochen took office and started his work immediately. In order to publicize and promote rural civilian education, he did the speech of *Why Do We Run Rural Civilian Education*? In the speech he said, "China's future, depends on the majority of the rural people. They are strong, China is strong; they are rich, China is rich; they are weak, China is weak; they are poor, China is poor." "If we want to transform China, the first step is to improve the people's wisdom and popularize education." Then Dr. Fu Baochen spent nearly a year editing and publishing *The Farmers' Textbook* and *Farmers Weekly*.

1928年，傅葆琛离开了平教会，到燕京大学、清华大学、北平师范大学等高等院校从事教育方面的教学工作。

In 1928, Dr. Fu Baochen left the Chinese Civilian Education Promotion Association and went to work in universities such as Yenching University, Tsinghua University, Peking Normal University in education.

1936年，傅葆琛回到了阔别40多年的家乡四川。此时四川省政府邀请晏阳初来四川组织一个设计委员会，该委员会分别设有教育、农业、卫生、地方行政4个专门的委员会，晏阳初再次请傅葆琛出马帮助他做教育计划方面的设计。为了规划四川未来的教育发展，傅葆琛到省内43个县、市进行深入调查。第二年，傅葆琛完成了调查工作，在此基础上撰写了《四川各县教育调查之经过及调查后发现之问题》调查报告。

In 1936, Dr. Fu Baochen returned to Sichuan, his hometown that he had been away 40 years. At this time, Sichuan Provincial Government invited Dr. Y. C. James Yen to organize a planning committee, which had Education, Agriculture, Health and Local Administration four special committees. Dr. Y. C. James Yen asked Dr. Fu Baochen again to help him to do education planning. In order to plan education of Sichuan, Dr. Fu Baochen went to 43 counties and cities in the province to conduct in depth investigation. The next year, Dr. Fu Baochen completed the investigation, and on this basis, he wrote the investigation report of *The Education Investigation of Sichuan Counties and the Problems Found in the Investigation*.

王锡林拍摄于20世纪50年代初的育德堂全景。王锡林提供。

Wang Xilin shot a panoramic view of the Cadbury School of Education Building in the early 1950s. Wang Xilin provided.

　　七七事变后，傅葆琛应聘到华西协合大学文学院担任教育组组长。从事了10多年乡村教育的傅葆琛发表了几十篇有关乡村教育的论文，形成了一套完整的乡村教育思想，他在国内成为以研究乡村教育而著名的学者。当时国内高等院校还没有乡村教育这样一个专业，因而傅葆琛就想在华西协合大学开设这样一个专业培养这方面的人才，让学生毕业后专门从事乡村教育。而此时教育部规定私立大学不得设教育系，校方经过教育部的特准，把原有的教育系改为乡村教育系，由傅葆琛出任系主任，该系是当时全国高校中唯一的一个关于乡村教育的系别。1944年奉教育部令，乡村教育系改为乡村建设系。

　　After the July 7th Incident, Dr. Fu Baochen was employed as the leader of the Education Group in the College of Liberal Arts, West China Union University. Dr. Fu Baochen had been engaged in rural education for more than ten years, published dozens of treatises on rural education, and formed a complete set of rural education thought. He became a famous scholar for studying rural education in China. At that time, there was no such a major in rural education in domestic colleges and universities. So Dr. Fu Baochen wanted to set up such a major in West China Union University to train talents in this field and specialize in rural education after graduation. At that time, the Ministry of Education stipulated that private universities were not allowed to set up education department. With the special approval of the Ministry of Education, the university changed the original Education Department to the Rural Education Department, with Dr. Fu Baochen as the department director. The Department was the only one among universities in China at that time. In 1944, the Department of Rural Education was changed to the Rural Construction Department by order of the Ministry of Education.

有了乡村建设系这样的舞台，傅葆琛更能施展他的才华。他出面主持新成立的教育研究所，创办《华西乡建》杂志。为了创办乡村建设学院，他与教务长蓝天鹤拜会了当时的西康省主席刘文辉，刘文辉当即捐款10亿国币用于修建乡村建设学院大楼，也就是教育学院大楼的西侧。

With the Department of Rural Construction, Dr. Fu Baochen showed his talents. He presided over the newly established Education Research Institute and founded *West China Rural Reconstruction* magazine. In order to establish the College of Rural Construction, he and provost Lan Tianhe visited Liu Wenhui, chairman of Xikang Province. Liu Wenhui immediately donated one billion Chinese currency to build the College of Rural Construction building, which was the west wing of the College of Education Building.

1949年，有朋友邀请傅葆琛到国外去，但他婉拒了。20世纪50年代初，由于乡村教育的理论不适应当时农村急剧变化的革命形势，乡村建设系停办了。傅葆琛离开了华西坝，到重庆西南师范学院教书去了。

In 1949, a friend invited Dr. Fu Baochen to go abroad, but he declined. At the beginning of the 1950s, the Rural Construction Department was closed down because the theory of rural education did not adapt to the rapidly changing revolutionary situation in the countryside at that time. Dr. Fu Baochen left Huaxiba to teach at Southwest Normal University in Chongqing.

金陵大学农学院对四川农业的贡献
Contribution of Agricultural College of University of Nanking to Sichuan Agriculture

抗战时期，金陵大学西迁到成都借助华西协合大学继续办学。当时该校迁到华西坝的院、系、科、所、部等单位很多，不可能全部安置在同一栋大楼里，只能分散在不同的教学楼里。金陵大学的行政部门就设置在育德堂里，这里是整个成都金陵大学的中心，学校里的重大决策、教务安排等都是在这里产生的。

During the Counter-Japanese War, University of Nanking moved west to Chengdu, and continued to run with the help of West China Union University. At that time, there were many colleges, departments and offices of the University of Nanking moved to West China Union University, so it was impossible to locate all in one building, but scatter in different teaching buildings. The Administrative Department of the University of Nanking was located in the Cadbury School of Education Building, which was the center of University of Nanking in

Chengdu. Major decisions and educational arrangement of the university were made here.

金陵大学在成都办学期间对四川贡献最大的是在农业方面。金陵大学早在1914年就开办了农科，是我国历史最悠久的高等农业教育学府。它以美国纽约州康奈尔大学农学院为样板，实行"教学、研究、推广"三位一体的发展模式。当时抗战急需为前方将士提供大量的农业物资，金陵大学农学院因此更加注重在四川农业方面实行"教学、研究、推广"的活动。

When University of Nanking was in Chengdu, the greatest contribution it made to Sichuan was in agriculture. As early as 1914, the University of Nanking had set up an agricultural department, which is the oldest institution of higher agricultural education in China. It took the College of Agriculture of Cornell University in New York as a model, and implemented the trinity development mode of "teaching, research and promotion". At that time, there was an urgent need to provide a large number of agricultural materials for the soldiers on the front line of the Counter-Japanese War, so the College of Agriculture paid more attention to the implementation of "teaching, research and promotion" in Sichuan agriculture.

金陵大学在华西坝办学期间行政部门就安置在育德堂里。这是成都金陵大学人气最旺的地方，学生报到注册、交学费，老师领薪水，师生取信件都要到此处办理。打开育德堂大楼一楼两旁的办公室窗户，就是办事窗口，大楼宽大的台基正好可以容纳下来办事的同学、老师们。图片来源：耶鲁大学。

When University of Nanking was running in Huaxiba, the Administrative Department was placed in the Cadbury School of Education Building. This is the most popular place of University of Nanking in Chengdu, where students register, pay tuition fees, teachers receive salaries, and every one gets letters. To open the office windows on both sides of the first floor of the Cadbury School of Education Building, where are the office windows. The spacious platform of the building is enough to accommodate for students and teachers to handle affairs. Photo source: Yale University.

金陵大学农学院在成都办学时有农业经济学系、农艺学系、植物学系、森林学系、园艺学系、蚕桑学系、植物病虫害学系和农业教育学系8个本科系。这8个系在成都对四川水稻、小麦、棉花、柑橘和烟草等农作物种植进行了推广与深入的研究。

The College of Agriculture of University of Nanking had eight undergraduate departments in Chengdu, namely, Department of Agricultural Economics, Department of Agronomy, Department of Botany, Department of Forestry, Department of Horticulture, Department of Sericulture, Department of Plant Diseases and Insect Pests, and Department of Agricultural Education. These eight departments conducted extensive and in-depth research on Sichuan rice, wheat, cotton, citrus and tobacco crops in Chengdu.

王绶是作物育种学家，1931年赴美国康奈尔大学作物育种学系读研究生。1933年回国后历任金陵大学农学院教授、农艺系主任、农艺研究部主任。他长期从事大豆、大麦育种研究，早在1924年王绶等人就培育出了"金大332"大豆，该新品种被外国人称为"南京大豆"。金陵大学西迁成都后，王绶继续在四川大力推广"金大332"大豆的栽培。在温江县试验时，大豆亩产量比当地农家品种高出许多。王绶与马育华又继续利用四川的大豆品种进行试验研究，在成都先后育出4个纯系品种，有些品种的产量和蛋白质含量比"金大332"大豆还要高。

Dr. Wang Shou was a crop breeder who went to the Department of Crop Breeding at Cornell University as a graduate student in 1931. After returning to China in 1933, he served successively as professor of School of Agriculture, director of Agronomy Department and director of Agricultural Research Department in the University of Nanking. He has been engaged in soybean and barley breeding research for a long time. As early as 1924, Dr. Wang Shou and others cultivated "Jinda 332" soybean, a new variety and was called "Nanjing Soybean" by foreigners. After moving west to Chengdu, Wang Shou continued to vigorously promote the cultivation of "Jinda 332" soybeans in Sichuan. When they experimented in the Wenjiang County, the yield per mu was much higher than the local farm varieties. Dr. Wang Shou and Mr. Ma Yuhua continued to use the soybean varieties in Sichuan to experiment and bred four pure varieties in Chengdu, some of which had higher yield and protein content than that of "Jinda 332" soybean.

金陵大学农学院来成都办学后不久，包望敏等人在仁寿乡村了解到当地农民迫切需要农业教育的愿望。随后农学院开办了金陵大学农学院附设仁寿县煎茶溪农民初级学校，由包望敏兼任校长。这所农民补习学校招收受过小学教育而在家务农的年轻人。特别有意思的是，为了避免非农民进入该校学习，报考时还要甄别一下考生是否是农民，那就是让考生用大水桶挑水，凡挑得起水的考生才能参加笔试，第1期录取了50名农村青年参加。

Shortly after the School of Agriculture of the University of Nanking moved to Chengdu, Mr. Bao Wangmin and others learned about the urgent desire of local farmers to agricultural education in Renshou countryside. Later, the College of Agriculture of the University of Nanking set up an attached Renshou County Jian Chaxi Farmers Elementary School, with Mr. Bao Wangmin as the principal. The school enrolled young people who had a primary education and farmed at home. What was particularly interesting was that in order to avoid non-farmers entering the school to study, when people applied for entering examination, the school needed to screen whether the examinee was farmers. So the candidate must carry big bucket of water. The candidate who could carry water could take the written test. The first phase admitted 50 rural youth to participate.

1943年2月5日，金陵大学在华西坝举行了农学院建院30周年的纪念活动，四川省主席张群出席了庆典活动，张群在庆典会上说："（金陵大学）今因抗战迁川，之我四川近水楼台，获益尤多。例如在校川籍同学，大学本部占三分之一，农业专修科超过三分之一，研究所及各训练班，亦以川籍同学为最多。2905号小麦为四川农村最普遍栽培之品种……故金大农学院对国家之贡献固大，四川之贡献尤多。"

On February 5, 1943 in Huaxiba, the University of Nanking held a commemoration of the 30th anniversary of the establishment of the Agricultural College. Zhang Qun, chairman of Sichuan Province, attended the celebration. Zhang Qun said at the celebration, "Now(The University of Nanking) moved to Sichuan due to the Counter-Japanese War. Sichuan has benefited a lot. For example, students from Sichuan account for one third of the university headquarters, more than one third of the agricultural subjects, and the institute and each training class, are also the largest number of Sichuan students. No.2905 wheat is the most commonly cultivated variety in rural Sichuan…Therefore, The Agricultural College of University of Nanking has made great contribution to the country and especially to Sichuan."

中国电化教育的开山宗师孙明经
Sun Mingjing, the Founder of Education Technology in China

电影在其诞生的第二年即1896年就传入了中国，但主要用在娱乐方面，直到20世纪30年代电影才在教育方面展现了它的强大功能。然而那时中国尚无培养电影人才的高等院校，直到1938年，西迁四川的金陵大学的孙明经创办了电化教育专修科，才结束了这种局面。

Film was introduced to China in 1896, the second year of its birth, but it was mainly used for entertainment. It was not until the 1930s that film showed its powerful function in education. However, at that time, China did not have any colleges and universities to train film talents. Until 1938, Mr. Sun Mingjing with University of Nanking moved to Sichuan and founded the Department of Audio-Visual Education, which ended the situation.

孙明经1911年出生于南京一个影像世家，其父孙熹圣在山东登州文汇馆读书时就接触到电影，是最早使用电影机的中国人，是他把英文"CINEMA"翻译成"电影"的。

抗战期间,成都金陵大学在华西坝教育学院大楼里在国内首次开办了电化教育专修科,从而也推动了华西协合大学开展电化教育。这是金陵大学使用16毫米的电影放映机放映教学电影。图片来源:耶鲁大学。

During the Counter-Japanese War, the University of Nanking in Chengdu opened the specialized subject of Audio-Visual Education for the first time in China in the building of the College of Education Building in Huaxiba, thus also promoted the West China Union University to carry out Audio-Visual Education. This is University of Nanking using a 16mm movie projector to show teaching films. Photo source: Yale University.

Mr. Sun Mingjing was born in 1911 in Nanjing at a family of film artist. His father, Sun Xisheng, came into contact with movies when he was studying at Tengchow College in Dengzhou, Shandong Province. Sun Xisheng was the first Chinese to use the film machine and translated the English word "cinema" into Chinese "Dianying".

孙明经从小就接触电影,耳濡目染使他对电影产生了浓厚的兴趣。1927年,孙明经考上南京金陵大学。父亲带他去拜访了校长陈裕光,校长告诉他,"明经小弟欲求这个发展方向实在很好。今天中国无此专门学校和专业,正待将来由明经小弟开创"。鉴于国内当时还没有相关的专业,陈校长亲自为孙明经设计了要学的相关课程,从理科的"化工学""电机学""物理学",到文科的"国文""戏剧""音乐"等诸多课程。1934年,孙明经经过7年的学习才毕业留校任职,是我国第一位电化教育专业人才。

Mr. Sun Mingjing had been contact with movies since childhood, so he had a strong interest in movies. In 1927, He was admitted to the University of Nanking. His father took him to visit the president Chen Yuguang, who told him, "It's great that Mingjing desire this development direction. There is no such special school or major in China today.

Therefore, it needs to be created by Mingjing in the future." Since there were no relevant majors in China at that time, president Chen designed relevant courses for him including chemical engineering, electromechanics and physics in science, Chinese, drama and music in liberal arts. In 1934, after seven years of study, Sun Mingjing graduated and stayed in school. He was the first Audio-Visual Education professional in China.

1935年7月，孙明经策划并参与拍摄制作的纪实影片《农人之春》在比利时布鲁塞尔举行的农村电影国际比赛上获得特等奖第三名，这是中国电影第一次在国际电影节上获奖。此后两年里，孙明经受教育部和中国教育电影协会之约，拍摄了十余部地理风景电影。

In July, 1935, the documentary film *The Spring of Farmers*, planned and produced by Mr. Sun Mingjing, won the third prize in the Rural Film International Competition held in Brussels, Belgium, which was the first time that a Chinese film won an award in an international film festival. In the following two years, Sun Mingjing made more than 10 geographical scenery films appointed by the Ministry Education and the China Education Film Association.

1936年3月，国民政府发布了《国难时期教育方案》，在这一方案里确定了

1945年，为了向民众宣传防治肺结核病的科普知识，华西肺痨病疗养院院长罗光璧撰稿编写了《肺痨病防治法》幻灯片的稿本，成都金陵大学教育电影部由主任孙明经（站立者）负责组成了设计团队，这是他们在教育学院大楼的教研室里讨论设计方案。这套72张、解说时间15分钟的幻灯片由联合国影闻宣传处制作出来在全国300多个放映站放映。
图片来源：耶鲁大学。
In 1945, in order to promote the science knowledge of prevention and treatment of pulmonary tuberculosis, Luo Guangbi, President of Chengdu Huaxi Pulmonary Tuberculosis Sanatorium, wrote the *Prevention and Control of Pulmonary Tuberculosis* slide manuscript. The director of the Audio Visual Center of University of Nankinge in Chengdu Sun Mingjing (standing) is responsible for the design team. The photo shows they are discussing the design plan in the teaching and research section of the College of Education building. This set of 72 slides(Commentary for 15 minutes) was produced by the United Nations Film and Information Publicity Office. It was shown at more than 300 screening stations across the country.
Photo source: Yale University.

"推广播音教育"和"促进电影教育"，也就是"电化教育"，用现代科技手段来教育、唤醒民众的抗日激情。同年7月，金陵大学理学院教育电影部成立，院长魏学仁兼电影部主任，孙明经为副主任。随后，教育部和金陵大学联合举办教育部电化教育人员训练班，孙明经负责设计课程并担任"动片摄制""静片摄制""教育电影与摄制"三门课的教学。

In March, 1936, the National Government issued the *Education Program during the National Crisis* in which it decided to "promote broadcasting education" and "promote film education", that was "audio visual teaching", using modern scientific and technological means to educate and awaken people's Counter-Japanese passion. In July of the same year, the Audio Visual Center of Science College of University of Nanking was established, with the dean Mr. Wei Xueren as director of the Film Department, and Mr. Sun Mingjing as deputy director. Subsequently, the Ministry of Education and University of Nanking jointly held the Audio-Visual Education Personnel Training Class of the Ministry of Education. Mr. Sun Mingjie was responsible for designing the curriculum and teaching three courses, Moving Film Production, Quiet Film Production and Education Film and Production.

1937年7月7日，全面抗日战争爆发，金陵大学被迫西迁入川。金陵大学的校级行政部门、文学院、农学院以及理学院的大部分专业迁到成都华西坝，而孙明经所在的理学院教育电影部等单位则迁到重庆。当时沿海一带已被日本军队占领，海盐难以运出。为了让国人了解大后方四川的井盐，鼓舞军民士气，孙明经接受中国教育电影协会等的委派带领助手范厚勤到自贡拍摄了《自贡井盐》和《井盐工业》两部纪录影片。

On July 7, 1937, when Counter-Japanese War broke out in an all-round way, University of Nanking was forced to move west to Sichuan. Most of the university-level administrative departments, the College of Literature, the College of Agriculture, and most of the majors of the College of Science of University of Nanking moved to Huaxiba, Chengdu, while the Audio Visual Center of the College of Science, with Mr. Sun Mingjing, and other units moved to Chongqing. At that time, the coastal area had been occupied by the Japanese troops, and the sea salt was difficult to transport out. In order to let Chinese people understand the well salt in Sichuan in the rear area and encourage the morale of the army and the people, appointed by the China Education Film Association, Mr. Sun Mingjing and his assistant Fan Houqin went to Zigong and shot two documentaries, *Zigong Well Salt* and *Well Salt Industry*.

随着国民政府用电影来宣传抗战建国的需求次数不断增加，电化教育人才的需求量也不断增加。孙明经曾说："电影是记录和传播文化的媒介，电影是教育和建设的利器。"1938年，教育部电化教育人员训练班第3期结束后，在教育部的支持下，金陵大学开办了电化教育专修科。这是我国电影高等教育的肇始，由于该门学科涉及文科和理科，为了更好地教学，电化教育专修科搬到成都，被安置在华西协合大学的教育学院大楼里。

With the increasing need of the National Government using films to publicize Counter-Japanese War, and building the country, the demand for film education talents was also increasing. Mr. Sun Mingjing once said, "Film is the medium to record and spread culture, and film is a powerful weapon for education and construction." In 1938, after the third periods of the Audio-Visual Education Personnel Training Class of the Ministry of Education, with the support of the Ministry of Education, University of Nanking opened Audio-Visual Education Specialization Section. This is the beginning of higher film education in China. As the subject involves both liberal arts and science, for better teaching, Audio-Visual Education Specialization Section was moved to Chengdu and was housed in the College of Education Building of West China Union University.

1941年，孙明经结束了在美国一年的考察电影与播音教育的工作，带着他在美国筹款购买的全套当时世界上最先进的电影、播音、摄影器材及教具回到成都，使当时地处成都的金陵大学的电影教学设备达到了世界一流水平。回国后，孙明经出任电影与播音部和电化教育专修科的主任。

In 1941, Mr. Sun Mingjing finished a year of studying film and broadcasting education in the United States, and returned to Chengdu. He raised money in the United States and bought a full set of the world's most advanced film, broadcasting, photography equipment and teaching equipment back, so that the teaching equipment of University of Nanking located in Chengdu at that time reached the world-class level. After returning to China, Mr. Sun Mingjing served as the Director of the Department of Film and Broadcasting and Audio-Visual Education Specialization Section of University of Nanking.

第二年3月，一本由孙明经主编的《电影与播音》杂志在成都华西坝问世，这是中国第一本集电影、广播、电视、摄影为一体的综合性学术月刊，孙明经在刊物上发表了大量学术文章。

In March of the next year, a magazine named *Film & Radio*, edited by Mr. Sun Mingjing, was published in Huaxiba, Chengdu. It was the first Chinese comprehensive monthly academic magazine of film, radio, television and photography. Mr. Sun Mingjing published a large number of academic articles in the journal.

1941年，金陵大学理学院教育电影部（音影部）与华西协合大学理学院联合，每学期在华西坝校园里放映教学电影。每周星期三下午4点至5点，在化学楼阶梯教室放映教学电影。每学期提前安排各学科的教学电影片名与讲解的老师，学生根据自己的专业选择上课。有资料显示，1943年全年共放映43场教学电影，学生观看人数达4 572人次。

In 1941, the Audio Visual Center of the College of Science of University of Nanking, jointly with the College of Science of the West China Union University, played teaching films on the Huaxiba campus every semester. Teaching films were shown in the lecture theatre of the Chemistry Building from 4 pm to 5 pm every Wednesday. Each semester, the teaching film names and the teachers of explanation were arranged in advance, so that students could choose classes according to their majors. Data showed that in 1943, a total of 43 teaching films were played, and the number of students watched reached 4,572.

自从金陵大学迁到成都，影音部与华西协合大学每周五晚上在华西坝放露天电影已经成为成都市的一项重要文化活动，每次都有近万名观众前来观看，孙明经时常还要现场解说。放映的内容不是故事片，而是包括我国在抗日战场上的进展、国际反法西斯战争的情况以及反映祖国的风光等纪录片。

Since University of Nanking moved to Chengdu, the open-air films in Huaxiba every Friday night showed by the Audio Visual Center and

抗战时期在华西坝每周五晚上都会在大操场上播放露天电影，它已经成为成都的一项重要文化活动。每次都有近万名观众来看电影，孙明经经常在现场解说。图片来源：耶鲁大学。
During the Counter-Japanese War, open-air movies were played on the playground every Friday night in Huaxiba. It had become an important cultural activity in Chengdu. Every time, nearly 10,000 audiences came to watch film, and Mr. Sun Mingjing often explained on the spot. Photo source: Yale University.

West China Union University had become an important cultural activity in Chengdu. Every time, nearly 10,000 audiences came to watch film, and Mr. Sun Mingjing often explained on the spot. The content is not feature film, but inclus the progress in China's Counter-Japanese battlefield, the situation of the international Anti-Fascist War, and documentaries reflecting the scenery of our motherland and so on.

1944年，孙明经受聘担任中国教育电影协会教课组副主任，负责设计筹划抗日战争胜利后全国电化教育的课程。抗战胜利后，孙明经随金陵大学返迁回南京，继续从事电化教育的教学工作。

In 1944, Mr. Sun Mingjing was hired as the Deputy Director of the Teaching Group of the China Education Film Association. He was responsible for the design and planning of the curriculum of the national Audio-Visual Education after the victory of the Counter-Japanese War. After the victory of the Counter-Japanese War, Mr. Sun Mingjing returned to Nanjing with University of Nanking to continue to work in the teaching of Audio-Visual Education.

而华西协合大学为了继续推进电影教学，1946年2月，经校长、校务长及文学院院长筹划，聘请校内4位教授成立了电影教育委员会，同时向英美两国订购电影教学器材，继续开展电化教育。

In order to continue to promote film teaching, in February, 1946, planned by the president of university, the dean of the university and the dean of the College of Arts, West China Union University invited four professors of the university to set up a Film Education Committee. Meanwhile, they ordered film teaching equipment from Britain and the United States to continue to carry out Audio-Visual Education.

华西坝老建筑的前世今生
The Stories of the Historic Architecture of Huaxiba

拾叁

西部医学教育的发祥地

启德堂

The Medical and Dental College Building,
the Cradle of West China Medical Education

1914年和1917年，华西协合大学先后分别开设了医科与牙科，此前按照荣杜易设计的校园总体规划，医学院的教学楼规划于校园中轴线旁东南方，但当时学校没能购买到这块地，只好在校园西北角置地修建医学院教学楼。

In 1914 and 1917, West China Union University opened Medical and Dental Departments respectively. According to the overall campus plan designed by Mr. Rowntree, the teaching building of the Medical College was planned to the southeast beside the central axis of the campus, but the university failed to buy the land, so it had to build a medical teaching building in the northwest corner of the campus.

1928年竣工的医学院教学楼由加拿大多伦多大学美道会所捐建，开始是建的东西相对的两幢建筑，东边为医科楼，西边为牙科楼，总称其为医牙科楼。20世纪40年代初修建了中间部分，把两头的医牙科楼连接成一栋楼。该楼的门廊前台阶是两层御路踏跺式石级阶，两边栏杆望柱上雕刻着龙，在台阶中间有分别代表中国文化的"阴阳八卦"和西医标志"蛇杖"的石刻图案。

The Medical College Teaching Building completed in 1928 was donated by Missionary Society of the Methodist Church in Canada of the University of Toronto. At the beginning, two buildings were built opposite to each other, the Medical Building in the east and the Dental Building in the west, generally known as the Medical and Dental Building. The middle section was built in the early 1940s to connect the Medical and Dental Buildings into one building. The front steps of the porch of the building are two layers of royal road stomping type stone steps, with loongs carved on the railing columns on both sides. In the middle of the steps there are stone carvings respectively representing the "Yin Yang Eight Trigrams" of Chinese culture and the "snake stick" of the symbol of western medicine.

由于医牙科楼在规划设计上的变动，当20世纪50年代成都市修建由北至南的人民南路时把大学分隔成东、西两部分。自此，位于西北角的医科教学楼自然便与学校主要建筑群分开了，之后该楼被命名为第八教学楼。

医牙学院教学大楼在20世纪40年代中期最终按照原设计图稿建成了。图片来源：王翰章提供，吉士道之子赠。

The Medical and Dental Building was finally built under the original design in the mid 1940s. Photo source: Wang Hanzhang provided, a gift from the son of H. J. Mullett.

医牙学院教学大楼一直是医科生、牙科生上课的教学楼,直到1964年在该楼的东边,新建的人民南路边修建了四川医学院附属口腔医院,牙科就全部搬迁到新口腔医院,现在该楼为华西临床医学院的教学楼。咸亚男摄于2021年。

The building of the Medical and Dental College was always the teaching building for medical and dental students. Until 1964, in the east of the building, the Dental Hospital Affiliated to Sichuan Medical College was built on the side of South Renmin Road and the Dental College moved to the new dental hospital. Now the building of Medical and Dental College is the teaching building of West China School of Medicine. Photo by Qi Yanan in 2021.

When the South Renmin Road from north to south was built in the 1950s in Chengdu, the university was divided into east and west two parts. Since then, the Medical Teaching Building in the northwest corner has naturally separated from the university's main complex, which was later named the Eighth Teaching Building.

20世纪60年代以后,第八教学楼台阶中间的"阴阳八卦"和"蛇杖"的石刻图案由频繁使用的台梯代替了,直到2005年在原址上重现石刻"八卦·蛇杖"图案,只是原来由红砂石修建的台阶与石刻换成了汉白玉。

After the 1960s, the stone carvings of "Yin Yang Eight Trigrams" and "snake stick" between the steps of the Eighth Teaching Building were replaced by steps. Until 2005, the stone carvings were reproduced on the original site. But the steps and stone carvings originally built of red sandstone have been replaced with white marble.

上图:1928年落成的医牙学院教学楼,左边为牙科楼(D),右边为医科楼(M),中间部分还没有建成。两栋楼的四周都是大片的农田。图片来源:四川大学档案馆。
The teaching building of the Medical and Dental College, completed in 1928. The Dental Building was on the left(D) and the Medical Building was on the right(M), and the middle part had not yet been built. Both buildings were surrounded by large areas of farmland. Photo source: Sichuan University Archives.

下图:这是一份建校初期华西协合大学医牙学院教学大楼的选址规划图,图中左边黑色E字形为医、牙科教学楼,E字形上面方形区域为医院。荣杜易最早规划设计的医牙学院的建筑在校园中轴线东侧,而这个选址规划图却完全远离了校园中轴线。这是因为大学校址这片区域是农田与墓地,购买这些土地的使用权不是一件容易的事情。赵安祝提供。
This is a site selection plan map of the teaching building of the Medical and Dental College of West China Union University in the early stage of the establishment of the university. On the left black E shape was the Medical and Dental Teaching Building, and the square area above the E shape was the hospital. According to the overall campus plan designed by Dr. Rowntree, the teaching building of the Medical College was planned to the southeast beside the central axis of the campus. But this site selection plan was completely far away from the central axis of the campus. This was because the university site was farmland and cemetery, and it was not easy to purchase the right to use the land. So the university had to build a medical teaching building in the northwest corner of the campus. Andrew George provided.

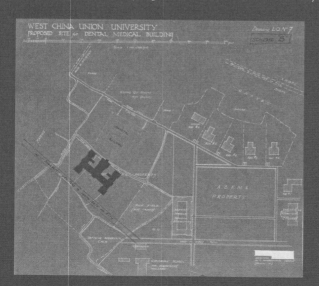

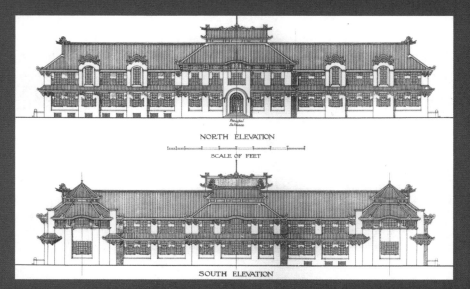

上图：1913年刊登在英国《建筑师》杂志上荣杜易设计的医牙学院教学楼的正立面图和背立面图。赵安祝提供。

The front elevation and back elevation drawings of the Medical College Teaching Building designed by Dr. Rowntree were published in the British magazine *The Builder* in 1913. Andrew George provided.

下图：1920年荣杜易设计的医牙学院教学楼的横截面和正立面图，从外形看比上图（1913年发表的）更简洁了。赵安祝提供。

The cross section and elevation drawings of the Medical College Teaching Building designed by Dr. Rowntree in 1920 are more concise in appearance than the above figure (published in 1913). Andrew George provided.

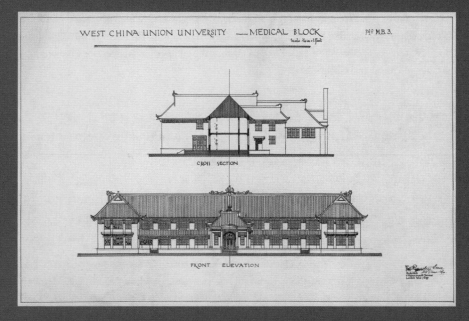

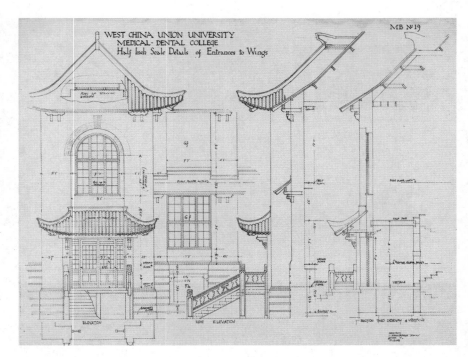

第八教学楼，现为四川大学华西临床医学院，2013年5月更名为启德堂。2020年7月13日，成都市人民政府将启德堂列为成都市第十七批历史建筑进行保护。

Now the Eighth Teaching Building is the location of West China School of Clinical Medicine, Sichuan University. The building was also named Qide Hall in May 2013. On July 13, 2020, the Qide Hall was confirmed as the 17th batch of Historical Buildings in Chengdu for Protection by the Chengdu Municipal Government.

中国西部现代医学高等教育的摇篮
The Birthplace of Modern Medical Higher Education in Western China

1910年，美国、英国和加拿大三国的五个基督教差会创办华西协合大学时，只开设了文科和理科，没有医科。加拿大差会的成员认为建医科，培养中国医生是很重要的，他们认为医学教育是传播基督教不可缺少的部分。因而在他们的努力下，1914年和1917年华西协合大学分别开设了医科与牙科。

In 1910, when five Christian missions from the USA, UK and Canada founded West China Union University, it opened only liberal arts and science, but no medical department. Members of the Canadian church considered to build medical department to train Chinese doctors was

荣杜易设计的医牙学院教学楼的两翼入口门廊图。赵安祝提供。

Entrance to wings diagram of the teaching building of the Medical and Dental College designed by Dr. Rowntree. Andrew George provided.

important. They thought medical mission as an indispensable part of spreading Christianity. Therefore, with their efforts, the university opened Medicine and Dentistry in 1914 and 1917 respectively.

然而办医科涉及诸多方面的问题，除了需要有基础学科和专业学科的教师外，还需要学生临床实习的医院等。建医科之初，学校只有一栋两层楼的临时用房和一些少量的实验仪器，教师有启尔德、莫尔思、甘来德和谢坚道4位医学博士，第一届招收了9名学生。

However, the establishment of Medical Departments involved many aspects. In addition to need teachers of basic and professional disciplines, it also needed students clinical practice hospital and other conditions. At the beginning of the Medical Department, there was only a two-story temporary room and a small number of experimental instruments. The teachers were Dr. Omar Leslie Kilborn, Dr. William Reginald Morse, Dr. Harry Lee Canright, and Dr. Charler Winfield Service, four MD. Nine students were enrolled in the first session.

他们遇到的第一个难题是医学教材，当时没有中文的医学教材，而内地成都学生的英文水平太低，不能像上海及北京等大城市的教会学校那样全用英文上课。更糟糕的是从上海购买的医学教科书在运输途中被水泡坏了，不能使用。外国教师们只好自己翻译教材，再请学校教国文的中国人抄写好，提供给学生用。

The first problem they encountered was medical textbooks. At that time, there were no Chinese medical textbooks, and the English level of mainland Chengdu students was too low to be taught entirely in English like church schools in big cities such as Shanghai and Beijing. To make matters worse, medical textbooks bought from Shanghai were damaged by water in transit and could not be used. The foreign teachers had to translate the textbooks themselves and then asked the Chinese teachers at the university to copy them and provide them to the students.

而医科学生的人体解剖课在当时是不被中国人接受的，中国人的传统观念是"身体发肤，受之父母，不敢毁伤，孝之始也"。在当时要想找一具尸体来上解剖课几乎不可能，不久发生了一起令人惊讶的事情，打破了这个局面。医学博士莫尔思曾回忆道：1914年的秋天，医科生刚入校。一天有人把一具无名的尸体放在了教室门口，这让他们很意外，不知道这是福还是祸，他们的确需要尸体上解剖课，但不是当时，而是要等先上一些基础课后才上解剖课。面对这具来历不明的尸体，他们不敢轻举妄动，害怕动了这具尸体引来祸事，但又不甘心放弃，这么容易就得来的尸体，要珍惜这难得的机会，上解剖课就不会纸上谈兵了。最后他们向学校汇报了此事，让校方向省政府请示，若确认为无人认领的尸体，请批准他们用来给学生上解剖课。得到政府方面的同意后，莫尔思改变了教学计划，提前给学生上解剖课。同时他们还建议校方邀请社会名流和政府官员前来参观他们上解剖课，消除社会民众对解剖尸体所产生的抵触情绪。

In addition, the human anatomy was not accepted by the Chinese people at that time. The traditional Chinese concept was that "The body given by parents, including hair and skin, should not be damaged, to protect the body is filial piety to parents". In those days it was almost impossible to find a corpse for anatomy class, and then something amazing happened that broke the situation. Dr. Morse once recalled the story. In the fall of 1914, medical students were just enrolled. One day someone put an unknown body at the door of the classroom, which surprised them. They did not know whether this was a blessing or a curse. They do need a corpse for autopsy class, but not now. They still had to wait for taking some basic class before taking the anatomy class. In the face of this unexpected harvest, they dared not act rashly, afraid of moving the body to cause trouble, but they were not willing to give up. The body came so easily. They should cherish this rare opportunity. The anatomy class would not be empty talking. Finally, they reported the incident to the university and asked the university to ask the provincial government for permission to use it to teach anatomy lessons if it was an unclaimed body. With the consent of the government, Dr. Morse changed the teaching schedule and started anatomy lessons earlier. They also suggested that universities invite celebrities and government officials to visit them for anatomy classes to overcome public resistance to dissecting corpses.

艰苦的日子终于熬过去了，1928年4月10日，为医科和牙科修建的永久性建筑医牙科楼迎来了剪彩的那一天。经过不断努力，医科和牙科培养了一批又一批的学生，这些学生在医牙学院的培养下成绩优异，在社会上有良好的口碑。1922年，美国纽约州立高等教育委员会同意，华西协合大学医科和牙科的毕业生，将其学业成绩送到该委员会复核后，就可获得该委员会的博士文凭。1920年，有4位医科学生毕业，此后每年都有医科学生毕业，到1952年，共有548名医科学生、188名牙科学生毕业。他们都分别获得中国华西协合大学和美国纽约州立高等教育委员会的医学博士学位。

The hard times finally passed, and on April 10, 1928, the permanent building for medicine and dentistry ushered in the ribbon cutting day. Through continuous efforts, the Medical and Dental College have cultivated a number of students, who have achieved excellent results under the training of the Medical and Dental College and have a good reputation in the community. In 1922, the University of the State of New York, then New York State Board of Education agreed that medical and dental graduates of the West China Union University could obtain the doctor's degree of the New York State Board of Education, after they submitted their academic achievement to the

1945年之后，医牙学院大楼门前台阶改建成两层御路踏跺式石级阶，两边栏杆望柱上雕刻着龙，在台阶中间有分别代表中国文化"阴阳八卦"和西医的标志"蛇杖"的石刻图案。照片中穿着白大褂的学生是华西协合大学医学院即将毕业的第35届医科学生。温良提供。

After 1945, the entrance steps of the Medical and Dental College Building were converted into two layers of royal road stomping type stone steps with loongs carved on the railing columns on both sides. In the middle of the steps there are stone carvings respectively representing the "Yin Yang Eight Trigrams" of Chinese culture and the "snake stick" of the symbol of western medicine. The students in the white coat in the photo are the 35th medical student of West China Union University Medical College who are about to graduate. Provided by Wen Liang.

Board for review. In 1920, four medical students graduated, and students graduated every year since. Until 1952, there were 548 medical students and 188 dental students graduated. They all received their MD degrees from West China Union University and New York State Board of Education respectively.

中国口腔医学之父林则
Ashley W. Lindsay, the Father of Chinese Stomatology

当年华西的医科和牙科联合组成学院，称为医牙学院。华西的医科在社会上的声誉很高，而华西的牙科是我国口腔医学的发源地。加拿大人林则是第一位由教会派到中国来的牙医传教士。林则是一位杰出的牙科医生、一位优秀的教师、一位高明的策划者、一位成功的创造者、一位能力极强的管理者，但

中国现代口腔医学创始人林则博士塑像。雕像和后面的博物馆建于2015年。该博物馆是四川大学华西口腔医院和成都市政府共同建立的国内首家口腔健康教育博物馆。该馆就建在华西口腔医院旁，距离地铁1号线华西坝站A出口30米，非常便于市民前去参观学习。咸亚男摄于2022年。Statue of Dr. Lindsay, founder of modern stomatology in China. The statue and the museum behind it were built in 2015. It is the first oral health education museum in China jointly established by West China Stomatological Hospital of Sichuan University and Chengdu Municipal Government. The museum is built next to West China Stomatological Hospital and 30 meters away from exit A of Huaxiba Station of Metro Line 1, which is very convenient for citizens to visit and study. Photo by Qi Yanan in 2022.

是，比上述更突出的是：他是一个有雄心壮志的人。

At that time, medical and dental department of West China Union University jointly formed college, called the Medical and Dental College. The Medical Department of the university has a high reputation in the society, and the Dental Department of the university is the birthplace of stomatology in China. In 1907, the Canadian Dr. Ashley W. Lindsay, served as the first dental missionary sent to China by church. Dr. Lindsay was a good dental practitioner, a good teacher, a brilliant planner, a successful creator, a very capable manager; but, more prominent than the above, he was an ambitious person.

林则1884年2月24日出生在加拿大魁北克省。1900年，16岁的林则高中毕业后离开家乡进入了多伦多皇家牙学院学习。在林则读大学的时代，加拿大掀起了宗教复兴运动，出现了历史上大规模的海外传教高潮。大学里也兴起了"学生志愿海外传教运动"。多伦多皇家牙学院的中国西部华西传教会也在大学校园里开展活动，吸收在校大学生毕业后到中国参加海外传教。林则也就是在那时开始对传教会感兴趣，通过传教会开展的活动，林则了解到在地球的另一头有一个东方帝国——中国。但林则对传教不感兴趣，他只想到中国西部做牙医，帮助那里的民众治疗口腔疾病。1906年秋，林则向传教会委员会递交申请，要求在传教会的支持下去中国西部做牙医。然而，由于林则不愿意保证执行传教会给他指派的传教职责，委员会很快驳回了他的申请。林则并没有灰心，他仍然和传教会保持联系。传教会的一位干事建议林则做事可以灵活一点，依他的能力和才干，不要认为同教会签了誓约卡就不能实现自己的愿望，如果他以医学传教士的身份到中国西部去，同样也可以达到他想为当地百姓治疗口腔疾病的心愿。

Lindsay was born on February 24, 1884, in Quebec, Canada. In 1900, at the age of 16, he left hometown after graduating from high school to attend the Royal College of Dental Surgeons in Toronto. In his college days, Canada set off a religious revival movement and there was a large scale overseas missionary climax appeared in history. The "Student Volunteer Overseas Missionary Movement" had also emerged in universities. The West China Mission of Western China at the Royal College of Dental Surgeons in Toronto also conducted activities on the university campus to absorb college students to take part in overseas missions in China after graduation. It was then that Lindsay became interested in the mission, and by its activities, he learned that there was an Eastern Empire at the other end of the earth —China. But he was not interested in missionary and only wanted to work as a dentist in western China to help people relieve their oral diseases. In the autumn of 1906, he submitted an application to the Missionary Committee, asking to be a dentist in Western China with the support of the missionary church. However, as he was reluctant to guarantee the mission duties assigned to him by the mission, the commission quickly dismissed his application. Dr. Lindsay was not discouraged, and he remained in touch with the mission. A secretary of the missionary association suggested that he could be more flexible; According to his ability, he should not think that he signed a pledge with the church can not realize his own wishes; If he went to western China as a medical missionary, he could also achieve his wish to relieve oral diseases for the people.

1907年，经过努力争取，林则到中国西部来做牙医的机会终于来了。林则与传教会签了誓约卡后，再次递交申请给传教会委员会，要求以医学传教士的身份到中国西部来传教。当年秋天，林则接到派遣他到中国西部成都来任医学传教士的任命。

In 1907, after hard efforts, Dr. Lindsay got the chance to western China to be a dentist. After signing a pledge card with the mission, Dr. Lindsay again submitted his application to the Missionary Committee requesting to go western China to preach as a medical missionary. In the fall of that year, he received an assignment sending him to Chengdu, western China, as a medical missionary.

1907年秋天，作为第一个到中国的传教牙医，踌躇满志的林则离开加拿大，踏上来中国的旅程。抵达上海后，沿长江而上，经重庆，林则一行终于在1908年的3月10日下午到达成都。

In the fall of 1907, as the first missionary dentist to China, the aspiring Dr. Lindsay left Canada on a journey to China. After arriving in Shanghai, traveling up the Yangtze River and passing through Chongqing, he and his group finally arrived in Chengdu on the afternoon of March 10, 1908.

早在1892年加拿大传教士启尔德博士等人就在四圣祠北街建立了成都仁济医院（现在成都市第二人民医院的原址）。所以林则到成都后就在仁济医院设立牙症诊所。在启尔德的帮助下，牙症诊所就安置在仁济医院所在的大院子的一间小房子里。

As early as 1892, Canadian missionary Dr. Kilborn and others founded the C. M. M. General Hospital on Sishengci North Street (Now the original site of Chengdu Second People's Hospital). So Dr. Lindsay set up a dental clinic in the hospital. With Dr. Kilborn's help, the dental clinic was located in a small house in the large courtyard of the hospital.

在林则还在来成都的路上时，启尔德的一位中国老朋友就请启尔德为其女儿诊治严重的颌部疾患，当时启尔德告诉他的朋友，一个牙科医生正在来成都的路上，并建议他们等牙医林则来了再医治。林则到成都不久那父女俩就知道了，他们立刻去找启尔德帮忙。启尔德便拜托林则医治他的这位中国朋友的女儿。检查时林则发现她患了10年的牙槽脓肿，牙槽骨质破坏，充满脓液，病人既痛苦又无信心。幸运的是，通过拔牙和彻底的刮除术治疗，她的病情迅速得到改善，不久就康复了。这位病人是林则在成都的第一个中国病人。

While Dr. Lindsay was still on his way to Chengdu, an old Chinese friend of Dr. Kilborn invited him to treat his daughter with a serious jaw disease. Dr. Kilborn told his friend that a dentist was on his way to Chengdu and advised them to wait for the dentist Dr. Lindsay to come for treatment. Soon after Dr. Lindsay arrived in Chengdu, the father and daughter knew and they immediately went to Dr. Kilborn for help. Dr. Kilborn asked him to treat his Chinese friend's daughter. By the examination, Lindsay found that she had suffered from ten years of alveolar abscess. The alveolar bone was destructed and was full of pus. The patient was both painful and not confident. Fortunately, with tooth extraction and a thorough curettage, her condition improved quickly and she soon recovered. The patient was Dr. Lindsay's first Chinese patient in Chengdu.

由于当时林则的主要任务是学习中文，因此他要求那位病人和她的家人都不要对其他人提起他治疗她的事，但是她的康复事实使其很难保守秘密。因此，后来许多获知她康复消息的牙症病人都去找林则治疗。这样一来当地许多人都知道了仁济医院里有个牙症诊所，专门医牙痛，向林则求医的病人越来越多了。由于学习语言任务重且时间有限，林则不便抽太多时间给他们看病。

Since Dr. Lindsay was tasked at the time to learn Chinese, he asked both the patient and her family not to mention his treatment of her to anyone else, but the fact of her recovery made it difficult to keep her secret. As a result, many dental patients who learned of her miracle of recovery went to Dr. Lindsay for treatment. In this way, many local people knew that there was a dental clinic in the C. M. M. Hospital, specializing in toothache, and more and more patients were seeking medical care from Dr. Lindsay. Due to the heavy task of language learning and limited time, Dr. Lindsay was inconvenient to spend too much time to treat them.

林则到中国仅3年，他的牙医工作就得到了教会的认可，同时因为牙科的迅速发展，林则也需要有合作者来和他一起发展他的事业——在中国开展牙科教育，为中国培养牙医人才，不仅给西部地区的百姓治疗牙患，而且要给整个中国的百姓提供牙医服务。

After only three years in China, Dr. Lindsay's dental work was recognized by the church, and because of the rapid development of dentistry, he also needed to have collaborators come with him to develop his career — dental education and training dental talents for China, not only to provide dental services to the people of the western region, but also to people of whole China.

1911年，教会决定修建仁济牙症医院并把林则大学时的同学唐茂森牙医博士指派来成都做牙科医生。成都教育史上的另一件大事对林则在中国开展牙科事业也是至关重要的，那就是1910年在成都的南门外锦江河畔后来被称为华西坝的地方，美国、英国和加拿大三国的五个基督教差会开办了华西协合大学，这为将来成都牙医专业的大发展打下了良好的基础。

In 1911, the church decided to build the C. M. M. Dental Hospital and assign Dr. John E. Thompson, a dental classmate of Dr. Lindsay in college to Chengdu as a dental doctor. Another event in the history of Chengdu education was also crucial for Dr. Lindsay to carry out dental career in China. In 1910, five Christian missions from the USA, UK and Canada established the West China Union University, which laid a good foundation for the future development of dental profession. The university located by the Jinjiang River, outside the South Gate of Chengdu.

1914年，华西协合大学医科开办了，林则和唐茂森都受聘于该大学，为医科学生讲授牙科课程，同时仍主持仁济牙症医院的工作。为了在中国发展牙科事业，1917年林则又邀请加拿大吉士道牙医博士来成都华西坝，在华西协合大学建立起牙科系，由林则任系主任。1919年，牙科正式扩建为与医科并列的牙科学院，林则担任院长，这是中国最早建立的近代高等牙科教育专业，按西方近代牙医学模式培养我国高等牙科医师的牙科学院。刚成立的牙科学院还没有一位牙科学生。林则从医科三年级学生中选中了黄天启，说服他转学牙科，因此，1921年中国便有了自己的第一位牙科医生，这也是亚洲的第一位牙科毕业生。黄天启毕业后留校任教，1928年到加拿大多伦多大学牙学院进修，原计划安排学习两年，得益于他在华西牙科学院获得的广泛扎实的基础知识及丰富全面的专业技能，一年后他便获得牙医博士学位。1936年华西协合大学牙科学院培养了中国第一批牙医女博士张琼仙和黄端方。

In 1914, the Medical Department of West China Union University was opened, and both Dr. Lindsay and Dr. Thompson were employed at the university to teach dental courses to medical students, while still hosting the work in the dental hospital. In order to develop his dental career in China, Dr. Lindsay invited Dr. Harrison J. Mullett to Huaxiba, Chengdu, to establish the Dental Department at the West China Union University, with Dr. Lindsay as the head of the department. In 1919, dentistry was officially expanded to the Dental College juxtaposed with the Medical College and Lindsay served as the dean. This was the earliest modern higher dental education major established in China. It was a dental college that trains Chinese higher dentists according to the western modern dental model. In the newly established Dental College, there was not a dental student yet. Dr. Lindsay selected Huang Tianqi from the third grade medical students and persuaded him to transfer to dentistry. So in 1921, China had her first dental practitioner. He was also the first dental graduate in Asia. He taught at the university after graduation. In 1928, he went to the Dental College, University of Toronto, Canada for further study. He was originally scheduled to study for two years. Thanks to the extensive solid basic knowledge and rich and comprehensive professional skills acquired in Dental College of West China Union University, he was awarded the Doctor of Dental Surgery degree one year later. In 1936, the Dental College of West China Union University trained Dr. Zhang Qiongxian and Dr. Huang Duanfang, the first female dentists in China.

1936年第22届华西协合大学医牙学院牙科毕业生与老师合影。前排是6位毕业生，中间两位手牵着手的女士黄端方（左）、张琼仙（右）是中国最早的两位牙科女博士；其余毕业生有张乐天（左1）、夏耀珊（左2）、张书林（右1）、谢尔琪（右2，俄国人）。第二排老师有黄天启（左1，他是中国第一位牙科毕业生）、吉士道（左2）、刘延龄（右1）；最后一排左为教务长方叔轩，右为校长张凌高。图片来源：四川大学华西口腔医学院修复科提供。

Group photo of dental graduates and teachers of the 22nd West China Union University School of Medicine and Dentistry in 1936. The six dental graduates are in front row. The two women holding hands in the middle, Huang Duanfang (left) and Zhang Qiongxian (right), are two earliest female dental doctors in China. The other graduates are Zhang Letian (1st from left), Xia Yaoshan (2nd from left), Zhang Shulin (1st from right) and Eugenie Sharevich (2nd from right, Russian). The teachers in the second row are Huang Tianqi (first from the left, he is the first dental graduate in China), H. J. Mullett (second from the left), R. G. Agnew(first from the right). The last row, the dean Fang Shuxuan on the left the principal Zhang Linggao on the right. Photo source: Department of Prosthodontics, West China School of Stomatology, Sichuan University.

虽然远离西方，且当时的中国条件很差，但是林则他们并没有降低开办牙科学院的标准。他们的教育政策和教育水平都站在西方的牙科学校的前列。牙科学生首先要受过与临床医科学生相等的基本生物学和医学的训练，然后再学牙科各种专业课程。华西牙科不像一般牙科学校那样偏重技术，而是要学生认识口腔卫生的重要性及其与全身健康的关系，这为中国建立了近代牙医教育的基础。在学制上医科和牙科都比其他学科多3年，总共要学7年。在教学中牙科学生和医科学生一样要通过严格的考试制度实行淘汰制，只要有一或两门主科不及格，就不能升级继续学医科或牙科，而必须转学其他学科。自此后至今中国各地口腔医学人才大多数都是从华西毕业的学生。华西牙科能取得如此成就，且成为全国牙科医学中心，林则功不可没，其被称为中国"牙科学之父"。

Though they were far away from the West and the conditions of China was poor at that time, they did not lower the standards of running the Dental college. Their educational policy and educational level were in front row of western dental colleges. In the Dental College of West China Union University, students were first trained in basic biology and medicine equal to clinical medical students before taking various professional courses in dentistry. Different from general dental colleges which focus on technology, but the Dental college of West China Union University requires students to understand the importance of oral hygiene and its relationship with systemic health, and to establish the foundation for modern dental education in China. The length of schooling in both medicine and dentistry are three years longer than in other disciplines, and seven years to study. In teaching, dental students and medical students should passed a strict examination system to implement the elimination system. As long as one or two main subjects failed, they could not continue to study medicine or dentistry, and must transfer to other disciplines. Since then most of the stomatology talents of China have been graduates of West China Union University. The Dental College achieved such achievements and become a national dental center. Dr. Lindsay has made a great contribution and is known as China's "Father of Dentistry".

中国西部第一位医科毕业学生李义铭
Li Yiming, the First Batch of Medical Graduate in Western China

李义铭是中国西部第一批医科毕业学生。图片来源：四川大学档案馆。

Dr. Li Yiming was the first batch of medical graduate in western China.
Photo source: Sichuan University Archives.

1914年华西协合大学医科招收了第一届9位学生，到1920年有4位毕业了。在这4位医科毕业生中，最富有传奇色彩的当属李义铭。李义铭这个贫困的中国男孩在学业完成之后，放弃了在洋人医院做医生的优越待遇。通过自强不息的奋斗，他偿还了由教会资助的学费，并在重庆开办了自己的医院。

In 1914, the Medical Department of West China Union University enrolled nine students in the first session, and four graduated in 1920. Among the four medical graduates, Dr. Li Yiming was the most legendary. He was a poor Chinese boy, completed his education he was unwilling to serve the church, and gave up the superior treatment as a doctor in the foreigner's hospital. Through constant struggle for self-improvement, he repaid the tuition fees funded by the church and opened his own hospital in Chongqing.

李义铭1894年出生于四川省自贡市，父亲李焕章是一位木匠。清朝末年，传教士到自贡传教，

1915年中国西部地区第一批西医学生与教师合影。后排医科老师从右起：启尔德、莫尔思、甘来德、艾文、谢道坚。前排医科学生从右起：李义铭、刘月亭、胡承先等7人。
图片来源：《华西协合大学》1915—1916。
In 1915, the first group of western medicine students in western China took a photo with their teachers.
Medical teachers in the back row from right: Dr. O. L. Kilborn, Dr. W. R. Morse, Dr. H. H. Canright, Dr. H. W. Irwin, Dr. C. W. Service.
Medical students in the front row from right: Li Yiming, Liu Yueting, Hu Chengxian and others.
Photo source: *West China Union University* 1915-1916.

兴修教堂。李焕章到教堂做木匠活，勤劳朴实的他终日里只知流汗埋头苦做，不知道什么叫歇歇。洋牧师用夹生的中国话让他休息休息，他也只是点点头，笑一笑，依旧干他的活儿。洋牧师相中了老实木匠李焕章，教堂建好后留下他在教堂里打杂看门，后来李焕章做了采购日常生活物品的管事。

  Li Yiming was born in 1894 in Zigong City, Sichuan Province. His father, Li Huanzhang, was a carpenter. At the end of the Qing Dynasty, missionaries went to Zigong to preach and build churches. Li Huanzhang went to church to work as a carpenter. He was hardworking and simple. He worked hard all day and was unwilling to rest. The foreign priest in unskilled Chinese, let him rest. But he only humbly nodded, smiled, and continual did his work. The foreign priest took a fancy to his the honest character. After the church was built, the foreign priest left Li Huanzhang to do chores and guard the door in the church, later as the steward of the purchase of daily life items.

  李义铭15岁时进了学校的大门读书学习。因在教会学校上学，尽管可以不出学费读书，但生活费用还得李义铭家里解决。由于家境清贫，李义铭每天只吃一顿饭。这个从小就品尝了劳苦的孩子，深知读书机会的来之不易，他确信必须要靠他一个人奋斗出来，他的家庭才可能彻底摆脱贫困处境。勤奋而聪慧的他刻苦努力地学习，只用5年的时间就读完了小学和中学的课程。

  At the age of 15, Li Yiming entered the school to study. Although he was able to study in church run school without paying tuition, his family had to pay for his living expenses. As his family was poor, he only ate one meal a day. The child, who had tasted the hard since childhood, knew that the opportunity to study was not easy. He was sure that he had to work hard to get his family out of poverty. Diligent and intelligent, he studied hard, finishing primary and secondary school courses in only five years.

  1914年，教会又出资让李义铭到成都华西协合大学读大学。在华西坝学习了6年后，李义铭于1920年毕业。根据规定，学生在教会的资助下毕业后，必须为教会服务5年。李义铭被派到重庆教会开办的宽仁医院当医生，每月的收入60银圆，这在当时是不少的薪金。但李义铭却放弃这份工作，想自己单独开家医院，一来通过自己行医、办实业来帮助他众多的兄弟姐妹，让父母能过上好日子；二来实现他的人生价值——"本医生天职为国家社会服务，冀渡贫苦病人于康乐之境"。

  In 1914, the church funded Li Yiming to attend West China Union University in Chengdu. After six years studying at the University, he graduated in 1920. According to the rules, students who church funded must serve the church for five years after graduation. Dr. Li Yiming was sent to work as a doctor in Kuanren Hospital, which was run by the Chongqing Church. He was paid sixty silver dollars a month, which was quite a salary at that time. But Dr. Li Yiming gave up this job and wanted himself to open a hospital alone. By practicing medicine, running industry, first he could help his many brothers and sisters, his parents could live a good life, second he could realize the value of his life —"As a doctor, my duty is to serve the country and society, so as to help the poor patients keep healthy and happy."

但是按照规定，由教会资助的学生毕业后没有为教会服务5年而离开教会医院，就必须偿还教会资助的2 000多银圆的学习费。李义铭筹措了一大笔款才偿还了教会资助的学费。他先在重庆打铁街办起了医务训练班，而后又在小什字街开设了义林医院。

However, according to the regulation of church, students funded by the church after graduation without serving the church for 5 years, and left the church hospital, must repay the church with over 2,000 silver dollar of study fees. Dr. Li Yiming raised such a large sum to repay his church-sponsored tuition. He first set up a medical training class in Chongqing Datie Street, and then set up Yilin Hospital in Xiaoshizi Street.

经过李义铭的精心经营，义林医院规模越来越大。1935年，为了扩展事业，他与重庆市市长张必果等组织了一个田园会来筹集资金。他在城郊观音岩购置土地筹建义林医院的新址。新建的医院大楼可谓倾注了李义铭大量的心血，从建筑的设计到施工他都事必躬亲。大楼的设计借鉴了李义铭母校华西协合大学内的中西合璧式建筑，也很巧妙地把钟楼设计在整幢大楼里。李义铭亲自监修，经过两年的修建，一幢坐北朝南的，具有大屋顶、青砖黑瓦、斗拱飞檐的中西合璧式大楼拔地而起，这幢7层的大楼在当时是重庆两个最高的建筑之一。

With the careful management of Dr. Li Yiming, Yilin Hospital was getting bigger and bigger. To expand his business, in 1935, he organized a small random loan association with Chongqing Mayor Zhang Biguo and others to raise funds, and he bought land in Guanyinyan, a suburb of the city, to build a new building for Yilin Hospital. Dr. Li Yiming poured a lot of effort into the new hospital building. From the design to the construction of the building, he attend to everything personally. The design of the building drew on the Chinese-Western combination architecture of the West China Union University, his alma mater, and the clock tower is also cleverly designed in the whole building. Dr. Li Yiming personally supervised the construction. After two years of construction, a south-facing Chinese and Western style combination building with a large roof, black brick tiles and bucket arch eaves was constructed. The seven-story building was one of the two tallest buildings in Chongqing at that time.

李义铭新建的义林医院大楼有5 000多平方米，除了很少的一部分是办公和居家用房外，其余大部分是医院用房。在李义铭全家刚入住大楼，并筹备医院开业时，全面抗日战争爆发了。1937年11月，国民政府宣布从南京迁都重庆。李义铭修建的义林医院大楼被国民政府征用，作为立法院、司法院及蒙藏委员会的办公大楼。一直到抗战胜利后，这幢大楼才退还给医院。

The new Yilin Hospital building covered more than 5,000 square meters, and most of it was occupied by hospital buildings, except for a small portion for offices and homes. When Dr. Li Yiming's family moved into the building and prepared to open the hospital, the Counter-Japanese War broke out in an all-round way. In November, 1937, the National Government announced to move the capital from Nanjing to Chongqing. Yilin Hospital was requisitioned by the National

李义铭借鉴其母校华西协合大学的中西合璧式建筑设计建造的义林医院大楼于1937年落成,当时这幢七层的大楼是重庆两个最高建筑之一。现在义林医院是重庆市人民医院·中山院区。戚亚男摄于2010年。
In 1937, Li Yiming built the Yilin Hospital Building, drawing on the Chinese and Western architectural design of his alma mater, West China Union University. At that time, the seven storey building was one of the two tallest buildings in Chongqing. Now Yilin Hospital is Zhongshan Hospital District of Chongqing People's Hospital. Photo by Qi Yanan in 2010.

Government as the office buildings of the Legislative Institution, the Judicial Institution and the Mongolian and Tibetan Committee. It was not until the victory of the War that the building was returned to the hospital.

义林医院这幢大楼不仅见证了西部地区第一位中国西医医生行医行善的生涯,还见证了抗日战争时期全民共同抗日的艰难历程。

Yilin Hospital building not only witnessed the career of the first western medicine doctor of China in the western region to practice medicine and do good deeds, but also witnessed the difficult process of the whole people fighting joint together against Japan during the Counter-Japanese War.

红岩英烈江姐原名江竹筠,李义铭是她的三舅。由于家境贫寒,江竹筠和弟弟童年时期几乎有一半的时间都是随母亲在自贡城里的外婆家度过的。

The Red Crag heroine sister Jiang formerly known as Jiang Zhujun. Li Yiming is her uncle. Because of their poor family, Jiang Zhujun and her younger brother spent almost half of their childhood with their mother in their grandmother's home in Zigong City.

李义铭与一起合办孤儿院的基督教同仁刘子如、曾子唯等商量,设法保送江竹筠姐弟进了孤儿院附设小学读书,那年江竹筠刚满12岁。江竹筠像当年舅舅李义铭小时候进学堂那样很珍惜这来之不易的机会，如饥似渴地学习，第一学期连跳三级。当年李义铭办的学校里常有老师是中共地下党员，在小学里这些老师启蒙了江竹筠对革命的向往。

Dr. Li Yiming discussed with his Christian colleagues Liu Ziru, Zeng Ziwei and others, who jointly organized a orphanage, and recommended Jiang Zhujun and her younger brother to study in the orphanage attached primary school. Jiang Zhujun had just 12 years old that year. She cherished her hard-won opportunity to enter school like her uncle entered school when he was a child. She studied eagerly and jumped at three grades in the first semester. In those days, the school often teachers were underground members of the Communist Party of China, in the school these teachers enlightened Jiang Zhujun's yearning for revolution.

1936年夏天，16岁的江竹筠以优异的成绩从孤儿院小学毕业了。当时她很想找一份工作赚钱改善家里的生活，但她又特别想继续求学。在这两难时，李义铭对外甥女江竹筠说："哪能不读书？考！考南岸中学！"重庆市南岸中学是李义铭等人办的，学校就设在李义铭的义林医院里边。这年秋天，江竹筠以优异的成绩考入了南岸中学。

In the summer of 1936, at the age of 16, Jiang Zhujun graduated from the orphanage primary school with honors. At that time, she wanted to find a job to make money to improve her family life, but she also had a strong desire to continue her education. In this dilemma, Dr. Li Yiming said to his niece, "How can you not go to school? Take the exam! Take Nanan Middle School Entrance Examination!" Chongqing Nanan Middle School was run by Dr. Li Yiming and others, and the school was located in Yilin Hospital. In that autumn, Jiang Zhujun was admitted to Nan'an Middle School with excellent grades.

在中学里江竹筠没有辜负舅舅李义铭对她的关爱，她努力学习。不久江竹筠获得了学校最高的奖励，一个嵌着镀银的"品学兼优"四个字的黑色盾牌。她把这奖品送给了舅舅李义铭。李义铭把这件礼物一直摆放在他办公室的书桌上，甚至后来江竹筠被捕入狱，他都没有把这件礼物从书桌上拿走。

In middle school, Jiang Zhujun did not disappoint her uncle. She studied hard. Soon she won the highest award of the school, a silver plated "Excellent in Character and Learning" black shield. She gave the prize to her uncle. Dr. Li Yiming kept the gift on the desk in his office. Even after Jiang Zhujun was arrested and put to prison, he didn't take it away from the desk.

在南岸中学读书期间，江竹筠开始在重庆寻觅党组织，向往加入共产党。时值抗日战争爆发，江竹筠积极参加学校组织的各种抗日活动，自己亲手缝棉衣，为前方战士募捐，到医院慰问伤病员。

While studying in Nanan middle school, Jiang Zhujun began to look for Party organizations in Chongqing and yearned to join the Communist Party. When the Counter-Japanese War broke out, she actively participated in various Counter-Japanese activities organized by the school. She sewed cotton padded clothes, to raised money for the soldiers in front, and went to the hospital to comfort the wounded and sick.

1939年春天，18岁的江竹筠考入校址在重庆远郊巴县的中国公学附属中学，成了高中部甲班的学生。在学校里江竹筠加入了共产党。

In the spring of 1939, at the age of 18, Jiang Zhujun was admitted to the Middle School Affiliated to China Public School in Baxian County, the outer suburb of Chongqing, and became a student of high school. At school, Jiang Zhujun joined the Communist Party.

江竹筠假期回舅舅李义铭家和同龄的表哥表弟表妹们在一起时，她常常把一些爱国的、革命的问题带来和大家一起讨论。他们谈起国家大事时常各执一词、激烈辩论。李义铭很为孩子们这样的争执担心，因为卫戍司令部就和他们同在一幢大楼里。但是他也制止不了她参加革命的热情，只得慨叹："我们家有小共产党了！"后来李义铭的二女儿被江竹筠引导参加了革命。

When Jiang Zhujun returned to her uncle house during the holiday to be with her cousins of the same age, she often brought some patriotic and revolutionary issues to discuss with everyone. When they talk about national affairs, they often hold their own opinions and debate fiercely. Dr. Li Yiming was very worried about the children's dispute, because the garrison headquarters were in the same building. He couldn't stop the revolutionary enthusiasm, so he had to sigh: "there is a small Communist Party in our family!" later, Li Yiming's second daughter was guided by Jiang Zhujun to participate in the revolution.

1946年暑期，江竹筠从国立四川大学回重庆。重庆地下党通知江竹筠不要回大学读书了，留在重庆搞革命工作。江竹筠立刻办了休学的手续，马上带上孩子彭云和丈夫彭咏梧到义林医院去看舅舅李义铭。舅舅李义铭得知江竹筠刚回重庆，还没有工作，他就主动建议外甥女在他管理的敬善中学做兼职会计。这个提议正合江竹筠的心意，这样她就有一个掩护身份的社会职业了。

In the summer vacation of 1946, Jiang Zhujun returned to Chongqing from the National Sichuan University. The Chongqing Underground Party informed Jiang Zhujun not to go back to university and stay in Chongqing to engage in revolutionary work. Jiang Zhujun immediately did the suspension procedures, and immediately took her son Peng Yun and her husband Peng Yongwu to the Yilin Hospital to visit her uncle. When Dr. Li Yiming learned that Jiang Zhujun had just returned to Chongqing and had no job, he suggested Jiang Zhujun to work as a part-time accountant in the Jingshan Middle School he managed. This proposal is in line with Jiang Zhujun's mind, so that she has a social profession to cover her identity.

1948年，江竹筠被捕入狱后，李义铭到处托人营救江竹筠，他还想用金钱把江竹筠保释出来，结果非但没有救出外甥女，他自己反而被国民党警备司令部抓去坐了好多天的牢。

In 1948, after Jiang Zhujun was arrested and put to prison, Dr. Li Yiming went everywhere to ask people to rescue Jiang Zhujun. He also wanted to bailout Jiang Zhujun with money. As a result instead of rescuing his niece, himself was caught by the Kuomintang police headquarter to put in prison for many days.

1951年，李义铭把义林医院捐给了人民政府，后来把医院大楼也交给了人民政府。重庆市卫生局在义林医院的基础上筹建了重庆市第二人民医院，李义铭任副院长。1956年，李义铭走完了他的一生。

In 1951, Dr. Li Yiming donated Yilin Hospital to the People's Government, and later handed the hospital building to the People's Government. Chongqing Municipal Health Bureau established Chongqing Second People's Hospital on the basis of Yilin Hospital, and Dr. Li Yiming served as Vice President. In 1956, Dr. Li Yiming died.

国内早期的生物化学研究所的创建人蓝天鹤
Lan Tianhe, the Founder of the Early Institute of Biochemistry in China

2006年，为纪念著名生物化学家蓝天鹤教授，四川大学华西校区内建立了一尊蓝天鹤铜像。威亚男摄于2022年。
In 2006, to commemorate the famous biochemist Professor Lan Tianhe, a bronze statue of Lan Tianhe was built in West China Campus of Sichuan University. Photo by Qi Yanan in 2022.

20世纪40年代初期华西协合大学医牙科楼修建了中间部分，把东西两端的医科楼与牙科楼连接成一栋楼，这部分建筑里除了有医牙学院行政办公室等用房外，还有学校的生物化学研究所，该所的创建者是华西协合大学生物化学教授蓝天鹤。

In the early 1940s, the middle part between of the Medical Building and the Dental Building of West China Union University was built. In this part of the building in addition to the administrative office of Medical and Dental College, there was also the university Biochemistry Institute. The founder of the institute was biochemist Lan Tianhe.

蓝天鹤1903年出生于四川荣县一个没落家庭，幼年丧母，从教会小学毕业后考入了成都华西协合中学。1927年蓝天鹤考入了山东齐鲁大学，在大学期间因参加抗日活动被追查，被迫转学至北平燕京大学化学系。1930年蓝天鹤从燕京大学毕业后，应聘到华西协合大学从事教学。之后蓝天鹤又到北平协和医学院进修生物化学，1936年回到华西协合大学，在医牙学院生物化学系从事教学工作，第二年他就担任系主任一职。

Lan Tianhe was born in a declining family in Rongxian County, Sichuan Province in 1903. He lost his mother when he was young. After graduating from primary school, he was admitted to Chengdu West China Union Middle School. In 1927, Lan Tianhe was admitted to Shandong Cheeloo University. When he was investigated for participating in Counter-Japanese activities and was forced to transfer to the Department of Chemistry of Yenching University in Peking. After graduating from Yenching University in 1930, Lan Tianhe was employed by West China Union University. After that, he went to Peking Union Medical College to study biochemistry. In 1936, he returned to West China Union University and taught in the Biochemistry Department of Medical and Dental College. In the next year, he served as the head of the department.

1940年，蓝天鹤获美国洛克菲勒基金会的奖学金，到美国纽约罗切斯特大学进修一年生物化学，并获得了博士学位。按照常理蓝天鹤应该回国了，但洛克菲勒基金会又提供经费让蓝天鹤再在美国做一年的研究工作。一年之后，蓝天鹤与国内家人的联系突然减少了，后来才知道是蓝天鹤应邀参加了美国的"曼哈顿"绝密计划，也就是研制原子弹的计划，该计划涉及物理、数学、化学、工程和医学等学科，有数十万人参加此项工作。蓝天鹤担任生物化学组副组长，生物化学属于医学部门，一共有研究人员200多人，该部门主要是研究原子能及α、β、γ射线对动植物以及人所产生的损伤。在两年多的时间里蓝天鹤只知道他在从事原子能的研究工作，直到1945年8月6日，美军对日本广岛投掷原子弹后，他才恍然大悟，原来他过去几年所从事的工作与制造原子弹有关。由此蓝天鹤对原子能的前途表示非常担心，他认为原子能在和平时期更应用于工业。

In 1940, Lan Tianhe was awarded a scholarship from the Rockefeller Foundation, USA, went to the University of Rochester in New York to study biochemistry for one year, and obtained a doctor's degree. According to common sense, Dr. Lan Tianhe should return home. However, the Rockefeller Foundation provided funds for him to do research in the United States for another year. A year later, the contact between Dr. Lan Tianhe and his family in China suddenly decreased. Later, it was learned that Dr. Lan Tianhe was invited to participate in the top secret Manhattan Project in the United States, that was, the plan to make atomic bomb. The project involved disciplines of physical mathematics, chemistry, engineering and medicine, and hundreds of thousands of people participated in this work. Dr. Lan Tianhe served as the deputy leader of the biochemistry team of medical department, with more than 200 researchers. The department mainly studied the damage of atomic energy, and α, β, γ rays to animals, plants and people. For more than two years, Dr. Lan Tianhe knew only that he was working on atomic energy. It was not until August 6, 1945, when the US military dropped an atomic bomb on Hiroshima, Japan, that he suddenly realized that their work in the past few years was related to make atomic bomb. Therefore, Dr. Lan Tianhe expressed great concern about the future of the atomic energy. He believed that the atomic energy should be developed for industrial use in peacetime.

1946年底，蓝天鹤回到国内，他于1948年在华西协合大学医牙学院创办了国内最早的生物化学研究所，并任所长，同时招收了6名研究生。其研究所分为生理化学、病理化学和食物化学三大部门，主要研究方向为放射化学、放射生物化学、组织化学、肿瘤生物化学、食物分析、病理食物及营养化学。该所就设立在医牙科楼刚修建的中间部分。蓝天鹤不仅把他回国时带回来的价值上万美元的仪器设备与药品捐献给研究所，而且还把他将去英国伯明翰大学担任客座教授第一年的薪酬600英镑全部用于研究所的建设。

At the end of 1946, Dr. Lan Tianhe returned to China. In 1948, he founded the earliest biochemistry institute in China at the Medical and Dental College of West China Union University, served as the director and enrolled six postgraduates. The institute was divided into three departments: Physiological Chemistry, Pathological Chemistry and Food Chemistry. Its main research aspects were radiochemistry, adiobiochemistry, histochemistry, tumor biochemistry, food analysis, pathological food and nutritional chemistry. The institute was located in the newly built middle part between of the Medical Building and Dental Building. Dr. Lan Tianhe donated to the Institute tens of thousands of dollars worth of instruments, equipment and chemical reagents that he brought back from abroad. He also devoted 600 pounds for the construction of the institute, which was his first year salary that he was going to a visiting professor at the University of Birmingham in the UK.

1949年11月，美国佛罗里达大学聘请蓝天鹤到美国讲学，就在他已经办好全家的签证和购买了机票准备前往美国前，他接受了中共地下党的建议放弃了到美国去的机会，决定留下来建设新中国。

In November, 1949, the University of Florida invited Dr. Lan Tianhe to give lectures in the United States. When he had his full family's visa signed and purchased air tickets to go to the United States, he accepted the suggestion of the Underground Party of the Communist Party of China, gave up the opportunity to go to the United States and stayed to build new China.

1950年，蓝天鹤等人随中国人民解放军西南军区卫生部入藏，发现并利用当地的野生植物沙棘，成功防治了坏血病（维生素C缺乏症）。1981年，已经78岁的蓝天鹤被批准为博士生导师，招收研究生。此后生物化学研究所从医牙科楼搬迁到基础医学院。

In 1950, Dr. Lan Tianhe and others entered Tibet with the Health Ministry of the Southwest Military Region of the Chinese People's Liberation Army, found and used the local wild plant sea-buckthorn to successfully prevent and cure scurvy (vitamin C deficiency). In 1981, the 78-year-old Lan Tianhe was approved as a doctoral supervisor to recruit postgraduates. Since then, the Institute of Biochemistry moved from the Medical and Dental Building to the Basic Medical College.

1988年，学校成立了生物化学与分子生物学研究所，蓝天鹤担任名誉所长，在成立大会上，他将自己的多年积蓄捐赠给研究所，建立了优秀研究生奖励基金。

In 1988, the university established the Institute of Biochemistry and Molecular Biology, and Dr. Lan Tianhe served as the Honorary Director. At the founding meeting, he donated his savings for many years to the Institute and established an Excellent Graduate Award Fund.

1991年11月23日，蓝天鹤去世，享年88岁。

Dr. Lan Tianhe died on November 23,1991, aged 88.

华西坝老建筑的前世今生
The Stories of the Historic Architecture of Huaxiba

拾肆

抗战催生的大学医院

华西协合大学医院

West China Union University Hospital, Which Was Built During the Counter-Japanese War

修建华西协合大学医院大楼是为了给华西协合大学医学院的学生以及抗战时期来华西坝联合办学的医学院的学生提供临床实习医院。大楼修建于1942年。该医院大楼位于华西坝的西面，医牙科大楼的北边，占地80余亩，有病床500张。医院大楼的外形呈凹字形，在长长的屋顶上除了有一水塔楼外，在屋顶拐角处有一个中式八角攒尖顶重檐的塔楼，大楼东边还建有瞭望塔。医院大楼与医牙学院大楼正好合围起来形成庞大的建筑群，堪称中国西部医学城堡。

Construction of the West China Union University Hospital building was to provide clinical practice hospital for Medical School students of West China Union University, and Medical School students who came to Huaxiba during the Counter-Japanese War. It was built in 1942. The hospital building was located in the west of Huaxiba, north of the Medical and Dental Building, covering an area of more than 80 mu. There were 500 hospital beds. The shape of the hospital building was concave, in addition to a water tower on the long roof, there was a Chinese octagonal spire tower at the corner of the roof, the east of the building also built watchtower. The hospital building and the Medical and Dental College building enclosed together to form a huge building complex, which could be called the medical castle in western China.

华西协合大学医院正门入口为四柱三间式门廊，进出台阶分别设在左右两边，中间有石栏杆，栏杆柱头上刻有龙。门廊里面墙上挂着两块牌子，左边为华西·齐鲁大学联合医院，右边为华西大学口腔病院。图片来源：耶鲁大学。

The main entrance of West China Union University Hospital had four columns and three gateways. The entry and exit steps were located on the left and right sides, with stone railings in the middle, and loongs carved on the railing capitals. There are two signs hanging on the inside wall of the porch, with the United Hospital of Huaxi Qilu University in Chengtu on the left and Dental Hospital of West China University on the right. Photo source: Yale University.

华西医院行政楼进出台阶已经做了修改，去掉了中间的石栏杆，可以从正面进出。右边的瞭望塔掩映在茂密的梧桐树后。戚亚男摄于2012年。
The entry and exit steps of the administrative building of West China Hospital have been modified by removing the stone railings in the middle, so people can go in and out from the front. The watchtower on the right is hidden behind the thick sycamore trees. Photo by Qi Yanan in 2012.

现在华西协合大学医院大楼为四川大学华西临床医学院和华西医院的行政楼，该楼是2012年四川省第三次全国文物普查的重要新发现，被列为近现代重要史迹及代表性建筑，专家称"华西协合（大学）医院建筑群中西合璧，富于特色，设计风格简洁明快，功能实用，达到了东、西方和谐与统一的美感，堪称中国建筑史上的奇葩，是研究中国近现代建筑的宝贵资料"。

Now the West China Union University Hospital building is the Administration Building of West China School of Medicine and West China Hospital of Sichuan University. The building was a major new discovery of the Third National Cultural Relics Census in Sichuan Province in 2012, and was listed as an Important Modern Historical Site and Representative Building. Some experts said, "West China Union University Hospital building complex combined Chinese and western elements, rich in characteristics, simple and lively design style, practical function, reached the east and western harmony and unified beauty. It is a wonder in the history of Chinese architecture, and is a valuable data of studying modern Chinese architecture."

2020年12月30日，华西医院老建筑群被成都市武侯区文化体育和旅游局列为"一般不可移动文物"并挂牌进行保护。

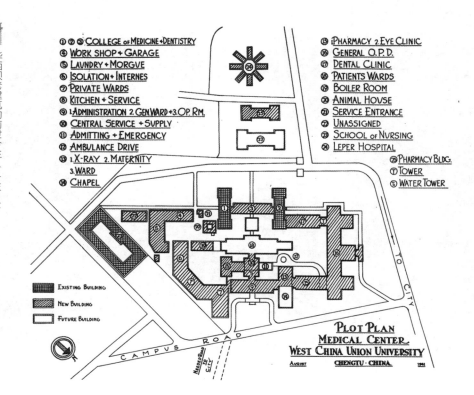

On December 30, 2020, the old building complex of West China Hospital were listed as "General Immovable Cultural Relics" by the Bureau of Culture, Sports and Tourism of Wuhou District, Chengdu and listed for protection.

华西医院的创始人之一启尔德
Omar Leslie Kilborn, One of the Founder of West China Hospital

华西协合大学医院的前身是由加拿大传教士、医学博士启尔德等于1892年在四圣祠北街建立的仁济医院。

The predecessor of West China Union University Hospital was the Gospel Hospital, also known as the C. M. M. Hospital, Renji Hospital, established by Dr. Omar Leslie Kilborn, a Canadian medical missionary and others in Sishengci North Street, Chengdu in 1892.

启尔德1867年11月20日生于加拿大安大略省一个铁匠的家庭，家里的5个孩子只有他和一个哥哥活了下来。小时候启尔德靠他哥哥挣钱读书，之后他凭打工赚钱继续读书，他做过铁路电话夜间接线生，还多次把水牛从加拿大运到英国贩卖。依靠打工的收入启尔德读完了高中和大学，他在皇后大学获得了化学硕士学位后，左图：1941年8月，华西协合大学医学中心规划总图。1. 医科楼；2. 牙科楼；3. 医牙学院；4. 车间、车库；5. 洗衣房、停尸房；6. 隔离室、实习生宿舍；7. 单人病房；8. 厨房、服务房；9. 行政楼、普通病房、手术室；10. 供应服务中心；11. 急诊室；12. 救护车车位；13. X光室、妇产科、病房；14. 小教堂；15. 医院药房、眼科门诊；16. 普通门诊；17. 牙科诊室；18. 病房；19. 锅炉房；20. 动物房；21. 服务入口；22. 未分配；23. 护士学校；24. 麻风病医院；25. 药房；T、塔楼；T1水塔
从这张规划图里可以看到一个拥有500张床位医院的规模有多大。图片来源：耶鲁大学。

Plot plan of West China Union University Medical Center, August 1941. This plan shows the size of a 500-bed hospital. Photo source: Yale University

下图：2013年，在成都第二人民医院门口设立了一座以"百年大爱"为主题的大型团体雕塑，展示了医务人员对人民生命的热爱。启尔德博士（站在后面左边）在雕塑中是西医进入四川的代表。戚亚男摄于2013年。

A large group sculpture with the theme of "One Hundred Years of Great Love" was set up at the gate of Chengdu Second People's Hospital in 2013, demonstrating the medical staff's love for people's lives. Dr. O. L. Kilborn (standing at back left)as the representative of western medicine into Sichuan was shaped in the sculpture. Photo by Qi Yanan in 2013.

22岁就获得了医学博士学位。

On November 20, 1867, Kilborn was born into a blacksmith's family in Ontario, Canada. Among the five children in the family, only he and his old brother survived. As a child, he relied on his brother to earn money to study. He then worked to earn money to continue studying. He worked as a railway telephone night operator and repeatedly transported buffalo from Canada to Britain for sale. Relying on his working income, he completed high school and college. After receiving his master's degree in chemistry from Queen's University, he received his doctor of medicine at the age of 22.

早在1889年启尔德还在学校读书时，他就作为一名基督教教徒给教会写信说他渴望做海外医学传教士的工作。1891年教会同意了启尔德的申请，决定由已在中国传教多年的赫斐秋带领启尔德、斯蒂文森两位医学传教士以及另外一位传教士来四川传教。同年年底，启尔德带着他的新婚妻子詹妮·福勒一行人离开加拿大，11月初抵达上海。启尔德他们在上海学习了3个月的汉语，次年2月16日，他们乘船进入四川，经过3个多月的行程于1892年5月21日到达了成都。

华西医院的创始人之一启尔德。
图片来源：启尔德外孙女黄玛丽提供。
Dr. Omar Leslie Kilborn, one of the founder of West China Hospital. Photo source: Dr. Kilborn's granddaughter Marion Walker provided.

As early as 1889, while still at university, Dr. Kilborn wrote to the church as a Christian saying that he was eager to work as a medical missionary overseas. In 1891, the Church agreed to Dr. Kilborn's application and decided that Dr. Virgil C. Hart (who had been preaching in China for many years), would lead Dr. Kilborn, Dr. D. W. Stevenson, two medical missionaries and another missionary came to Sichuan to preach. At the end of the same year, Dr. Kilborn left Canada with his newly married wife, Jennie Fowler Kilborn, a group of people arrived in Shanghai in early November. They studied Chinese in Shanghai for three months, and on February 16, the following year, they sailed into Sichuan, and reached Chengdu on May 21,1892.

启尔德他们乘坐的船抵达成都已经是傍晚6点钟左右，他们的出现立刻引起了大量市民的围观，之前有洋人来成都，那都是男的，但这次几位女洋人来，可是这座城市的头一回，围观那是必然的，他们一上岸就坐上轿子进城了。事后启尔德讲坐轿子就是两根竹竿绑在椅子两边，人坐在椅子上由两人肩扛而行。

When Dr. Kilborn and his group arrived in Chengdu at about 6 o'clock in the evening, their appearance immediately drew crowds of onlookers. There were foreigners came to Chengdu before, and all of them were men. But this time several female foreigners came, this was the first time of the city. Onlooking was inevitable. As soon as they got ashore, they took on the sedan chairs into the city afterwards. Dr. Kilborn said that the sedan chair was two bamboo sticks tied to both sides of the chair, and people sat on the chair and were carried by two peoples on shoulders to walk.

两周后赫斐秋在城里租了民房开始了他们从事医疗传教的工作。他们把租来的房子安排成一间诊疗室、一间可收治8到10人的病房、一间书房和一个用于传教的礼拜堂，余下的为他们自己居住的房间。他们一边加紧学习中文，一边准备传教工作的开展。就在他们信心十足准备开展工作时，厄运降临到他们身边，给了他们一个致命的打击。7月9日，詹妮·福勒生病了，根据她的症状启尔德判断他的妻子感染上了霍乱。第二天晚上，詹妮·福勒就去世了。由于霍乱持续在成都流行，安葬了詹妮·福勒后，他们只好暂时离开成都到几十公里以外的郊区山上躲避瘟疫。

Two weeks later, Dr. Hart rented houses in town and began their medical missionary work. They arranged the rented house into a treatment room, a ward for 8 to 10 people, a study and a chapel for missions, and the remaining room for themselves to live in. While they were learning Chinese, they were also preparing for missionary work. Just when they were confident to work, doom hit them and gave them a fatal blow. On July 9, Mrs. Kilborn became ill and Dr. Kilborn judged by her symptoms that his wife was infected with cholera. Mrs. Kilborn died the next night. As cholera continues to spread in Chengdu, after the burial of Mrs. Kilborn, they had to temporarily leave Chengdu to the suburban mountains dozens of kilometers away to avoid the plague.

启尔德一行在山上待到9月份才回到了成都，继续为开诊做准备。1892年11月3日，成都第一家西医诊所开诊了。这一天是启尔德他们到达中国一周年的日子，他们精心安排诊所在值得纪念的这一天开业。开业当天就有18位患者前来看病，随后前来看病的市民越来越多。

Dr. Kilborn and his group stayed in the mountains until September and returned to Chengdu to continue to prepare for the opening of the clinic. On November 3, 1892, the first western medicine clinic in Chengdu opened. This day is the first anniversary of their arrival in China. They carefully arranged the opening of the clinic on this memorable day. On the opening day, 18 patients came to see the doctor, and then more and more citizens came to see the doctor.

第二年的2月，加拿大教会又派遣了几位女传教士来四川，她们到达上海后，就一边学中文，一边等四川传教站派人去接她们。9月，启尔德被派到上海去接这几位女士来四川。在来四川的漫长路途里，启尔德与她们中一位名叫瑞塔的多伦多大学女子医学院毕业的学生相爱了。经过5个月的行程，启尔德带领这几位女士来到了成都。启尔德也与瑞塔结为了夫妻，他们俩作为一对搭档被派遣到乐山负责在那里开办一个医疗诊所。同年，夫妇俩的第一个孩子启真道在乐山出生了，这个孩子后来成长为华西协合大学医牙学院的院长。

In February of the following year, the Church of Canada sent several more female missionaries to Sichuan. When they arrived in Shanghai, they learned Chinese while waiting for Sichuan missionary station to send some one to pick them up. In September, Dr. Kilborn was sent to Shanghai to bring the female missionaries to Sichuan. During the long journey to Sichuan, Dr. Kilborn fell in love with one of them, named Retta Gifford, who graduated from the University of Toronto Women's School of Medicine. After a five-month journey, Dr. Kilborn led the missionaries to Chengdu. Dr. Kilborn also married Retta Gifford, and two were sent as partners to Le Shan in charge to open a medical clinic there, and in the same year the couple's first child, L. G. Kilborn was born in Leshan, who later grew up to become President of the Medical and Dental College of West China Union University.

不久，赫斐秋在四圣祠北街购买地产，修建了一所医院、一所教堂和一所学校。1895年4月，启尔德夫妇也被调回成都开展工作。后夫妇俩因一些意外原因返

回上海，并滞留到第二年4月才再次回到成都。

Soon, Dr. Hart purchased property and built a hospital, a church, and a school at Sishengci North Street. In April, 1895, Dr. Kilborn couple were also transferred back to Chengdu for work. Later some unexpected things happened. Dr. Kilborn couple with their children had to go to Shanghai and didn't return to Chengdu until April of the second year.

在随后的日子里，启尔德参与了创建华西协合大学的工作并曾担任该大学的理事会主席一职，他除了诊治病人外还在大学里讲授医学课程。1919年，启尔德夫妇俩回加拿大休假，次年启尔德不幸因肺炎在加拿大去世，享年53岁。

In the days that followed, Dr. Kilborn participated in the establishment of West China Union University and served as chairman of the Council of the University. In addition to treating patients, he also taught medical courses at the university. In 1919, Dr. Kilborn couple returned to Canada for vacation, where he unfortunately died of pneumonia the following year at the age of 53.

尽管启尔德去世了，他的理想却在他的家族延续下去了。从启尔德1892年来到中国，到1952年其长子最后离开中国，60年里启尔德一家三代有8位成员为华西协合大学的教育事业而工作。启尔德的妻子启希贤在丈夫去世后就返回了成都，继续在医学院任教，并在仁济女医院工作长达12年。她在华西协合大学一直工作到70岁，于1933年退休回加拿大。1942年12月1日，启希贤在多伦多去世，享年79岁。

Although Dr. Kilborn died, his ideals lived on in his family. From Dr. Kilborn's arrival in China in 1892 to his eldest son's final departure in 1952, eight members of the three generations of Dr. Kilborn had worked for educational cause of West China Union University for 60 years. Mrs. Kilborn returned to Chengdu after her husband died, continuing to teach at the medical school and working at the C. M. M. Hospital for Women and Children 12 years. She worked in West China Union University until she was 70 and retired back to Canada in 1933. On December 1, 1942, she died in Toronto at the age of 79.

当初启尔德他们创办的四圣祠仁济医院后来一直都是华西协合大学的教学医院，直到1942年华西协合大学在华西坝建了新医院，仁济医院的大部分人员和设备也都搬迁到了新医院，仁济医院规模缩小，改为了慢性病医院。再后来华西协合大学

2012年华西医院院史陈列馆建在八角楼下面。戚亚男摄于2021年。

In 2012, the History Museum of West China Hospital was established under the Octagon Building. Photo by Qi Yanan in 2021.

医院更名为四川医学院附属医院,即今天的四川大学华西医院,而仁济慢性病医院也就是现在的成都市第二人民医院。

The C. M. M. General Hospital founded by Dr. Kilborn and others at Sishengci North Street also had been the teaching hospital of West China Union University. Until 1942, when the university built a new hospital in Huaxiba, most of the personnel and equipment of the C. M. M. General Hospital were also moved to the new hospital. The scale of the C. M. M. General Hospital was reduced and changed to a chronic disease hospital. Later, West China Union University Hospital was renamed the Affiliated Hospital of Sichuan Medical College, now West China Hospital of Sichuan University, and the Chronic Disease Hospital is now the Chengdu Second People's Hospital.

尽管20世纪50年代初西方传教士都离开了中国,但启尔德家族与华西协合大学的交往并没有停止。1998年启尔德家族捐款建立了一个以启尔德命名的访问学者基金,该基金每1~2年根据华西医院的需求,选派一名医学家或者临床专家到华西医院讲课或临床交流2~3周。而项目的管理者之一罗伯特就是1923年出生于成都的启尔德的孙子、启真道的儿子。

Although all western missionaries left China in the early 1950s, the Kilborn family's relationship with the West China Union University did not stop. In 1998, the Kilborn family donated to establish a Visiting Scholars Fund named after Dr. Omar Leslie Kilborn. According to the needs of West China Hospital, every 1-2 years the fund sent a medical scientist or clinical expert to West China Hospital to make lectures or clinical communication for 2-3 weeks. Dr. Robert Kilborn, one of the project managers, born in Chengdu in 1923, was the son of Dr. L. G. Kilborn, the grandson of Dr. Omar Leslie Kilborn.

抗日战争催生的大学医院
The University Hospital Built during the Counter-Japanese War

自从1914年、1917年华西协合大学分别开设了医科和牙科后,教会便同意把在四圣祠北街开办的仁济医院和仁济牙症医院作为医科和牙科的临床实习医院。当时校方规划是要在学校里建立一所可供医科、牙科学生实习的医院,只是由于在总体规划里,设计用于修建医院的那块坟地始终没有能够从土地所有者手中购买到,加上当时学生可以到仁济医院去实习,因而修建大学附属医院的工程就一拖再拖。最后,校方决定放弃原来校园设计的规划,在校园西北面购买了一块土地用于修建大学医院。由于校方想要建一所当时国内一流的大学医院,因而一直在筹措资金,等待时机成熟时修建。

Since the West China Union University opened its Medical and Dental Departments in 1914 and 1917 respectively, the Church had agreed to make the C. M. M. General Hospital and Dental Hospital at Sishengci North Street as clinical practice hospitals for medical and dental student.

正在修建的华西协合大学医院，右边屋檐为医学院大楼。图片来源：四川大学华西医学中心提供，芳卫廉的女儿赠。

West China Union University Hospital under construction. The right is eaves of the Medical College building. Photo source: The West China Medical Center, Sichuan University, a gift from William P. Fenn's daughter (Mary Francis Hazeltine).

At that time, the university planed to build a hospital for medical and dental students to practice. Just because in the overall plan, the hospital was designed to build on a graveyard, but the university had not been able to buy the graveyard from the land owner, also students could go to the C. M. M. Hospital to practice, thus to build a university affiliated hospital of the project was delayed. Finally, the university decided to abandon the original campus design plan, and purchased a piece of land in the west of the campus for the construction of the university hospital. As the university designed to build a first-class university hospital in China at that time, it had been raising funds and waiting for construction when the time was ripe.

1938年，在华西坝联合办学的齐鲁大学和中央大学医学院为了让高年级的医科学生能进入临床实习，他们与华西协合大学经过协商，决定仁济医院、仁济牙症医院等医院不仅提供给华西协合大学学生实习用，而且以这几所医院为依托成立华西·中央·齐鲁三大学联合医院，由中央大学医院院长戚寿南担任总院长，三个大学的高年级医科学生都可以到这几所医院实习。

In 1938, Cheeloo University and National Central University School of Medicine were jointly running school in Huaxiba. In order to let senior medical students could enter clinical practice, after consultation with West

乌瞰落成后的华西协合大学医牙学院与大学医院，堪称中国西部医学城堡。图片来源：耶鲁大学。

A bird's eye view, the Medical and Dental College and the University Hospital of West China Union University can be called a medical castle in western China. Photo source: Yale University.

China Union University they decided the C. M. M. General Hospital, Dental Hospital and other hospitals not only provided West China Union University students internship, but relied on these hospitals, the United Hospital of the Associated Universities in Chengtu was established. Qi Shounan, the President of the National Central University hospital, served as the General Director. Senior medical students from the three universities could go to these hospitals for internship.

随着三大学联合医院的运行，越来越多的三校医科学生去临床实习，但城里四圣祠北街的仁济医院除了路途远不说，且也容纳不下这么多学生在那里实习。因而华西协合大学在华西坝建一所不仅可以提供给医科学生临床实习而且可为社会服务的医院就迫在眉睫。然而在战争时期要建一所综合性的医院是很不容易的，好在之前华西协合大学就筹有用于修建一所医院的资助经费7.5万元，初期的选址和一些辅助用房已经完成。随后校方经过多方面集资经费使得华西协合大学医院得以建成。远在美国的大学校务长毕启也为医院募集到200吨医疗物资，并在医院落成前由他亲自护送回华西坝。

With the running of the United Hospital, more and more medical students from the three universities went to clinical practice. But the C. M. M. Hospital in Sishengci North Street in the city was far away the

华西医学城堡在鳞次栉比的现代化高楼大厦中显得特别突出，该城堡现在是华西临床医学院与华西医院行政楼所在地。戚亚男摄于2022年。
The West China Medical Castle, standing out among the rows of modern high-rise buildings, where the West China Clinical Medical College and the West China Hospital Administration are located now. Photo by Qi Yanan in 2022.

university, and could not accommodate so many students to practice there. Therefore, it was urgent for West China Union University to build a hospital in Huaxiba that could not only provide clinical internship for medical students, but also served the society. However, it was very difficult to build a comprehensive hospital during the war. Fortunately, as early as 1936, West China Union University had previously raised 75,000 yuan for the construction of a hospital. The site selection in the early stage and building some auxiliary houses had been completed, and then the university raised funds in various ways to establish the hospital. The university President Dr. Beech, also raised 200 tons of medical supplies for the hospital from the United States, which he escorted to Huaxiba himself before the hospital was completed.

1941年，四川省卫生实验处与中央大学医学院在正府街联合建立成都公立医院，中央大学医科学生有了新的实习医院，随即退出了三大学联合医院。齐鲁和华西两个医学院仍继续实行合作，三大学联合医院因此改名为华西·齐鲁大学联合医院。

In 1941, Sichuan Provincial Health Experimental Department and Central University School of Medicine jointly established Chengdu Public Hospital in Zhengfu Street. Medical students of Central University School of Medicine had a new internship hospital and immediately withdrew from

华西协合大学结核病疗养院院长罗光壁（左）与华西·齐鲁大学联合医院院长杨济灵（右）在大学医院门前合影，他们背后的墙上挂着"华西·齐鲁大学联合医院"的牌子。图片来源：四川大学档案馆。

Dr. Luo Guangbi(left), President of the Tuberculosis Nursing Home of West China Union University, and Dr. A. E. Best(right), President of Huaxi and Qilu University United Hospital in Chengtu, took a photo in front of the university hospital. The sign of Huaxi and Qilu University United Hospital in Chengtu hung on the wall behind them. Photo source: Sichuan University Archives.

the United Hospital of the Associated Universities in Chengtu. Cheeloo University and West China Union University continued to cooperate, and the United Hospital of the Associated Universities in Chengtu was renamed Huaxi and Qilu University United Hospital.

同年，华西协合大学医院的口腔病医院率先建成并启用了。医院的内部设计很现代化，甚至与美国有名的牙病医院相比也毫不逊色。此外，医院的医疗器械也是最先进的，其新近从美国购买的新型X光机及齿髓检验仪等也正式装配启用。

In the same year, the Stomatological Hospital of the University Hospital took the lead in building and opening. The internal design of the hospital was modern, even compared with the famous dental hospitals in the United States. The medical equipments of the hospital were also the most advanced, and the new X-ray machine and tooth pulp detector recently purchased from the United States have also been formal assembled and enabled.

1942年，新建的大学医院部分已可投入使用了，医院新任副院长杨振华率先建立了两个外科病房供齐鲁大学和华西协合大学的医科学生临床实习。1943年4月10日，有500张床位的华西协合大学医院举行了开院典礼。

In 1942, part of the new university hospital was put into use. Dr. Yang Zhenhua, the new Vice President of the hospital, took the lead in establishing two surgical wards for clinical practice of medical students from Cheeloo University and West China Union University. On April 10, 1943, the 500-bed university hospital was inaugurated.

修建一所医院大楼容易，但要使它运行起来可就不那么容易了，除了要有医生外，最主要的是要有大量的护士。此时因为大学医院正缺护士，虽然医院有病房有设备，但不能开放更多的病房提供给教学和患者使用，这样就限制了该医院的正常发展。就在这时，校方得到了一条让人振奋的消息，北平协和医学院护士学校正准备在重庆恢复办学校。校方立刻与校长聂毓禅联系。

It is easy to build a hospital building, but it is not so easy to get it running. Besides doctors, the most important thing is to have a large number of nurses. At that time, the university hospital lacked nurses, and although the hospital had wards and equipment, they could not open more wards for teaching and serving patients, which limited the development of the hospital. At this time, the university received an exciting news that the Principal of Peking Union Medical College Nursing School, was preparing to resume running a nurse school in Chongqing. The university immediately contacted with the president, Ms. Nie Yuchen.

华西协合大学医院建筑群局部，左边为医院门诊部，中间为瞭望塔，右边为口腔病院新增建的诊室。图片来源：耶鲁大学。
Part of the West China Union University Hospital building complex, on the left was the Outpatient Department of the hospital, on the middle was the Observation Tower and on the right was new extension Dental Clinic. Photo source: Yale University.

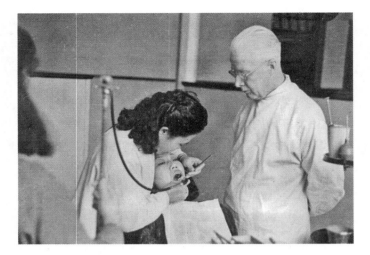

牙科博士古士道在牙科医院指导学生给牙病患者治疗。图片来源：耶鲁大学。

Dr. H. J. Mullett is instructing student to treat patient with dental diseases in the dental hospital. Photo source: Yale University.

聂毓禅1927年毕业于北京协和医学院护士学校，留校任医院病房副护士长。从1929年起她先后到加拿大多伦多大学医学院公共卫生系、美国哥伦比亚大学师范学院进修。1940年回国后，聂毓禅担任当时国内唯一的高等护理学校——北平协和医学院护士学校校长。1941年12月8日，日本偷袭珍珠港，美国对日宣战，随后协和护士学校被迫关闭。1943年春，聂毓禅带领协和护士学校的老师内迁，希望在大后方继续办护士学校。聂毓禅他们从沦陷区的北平经过两个多月艰难旅途才到达重庆。聂毓禅在筹备协和护士学校恢复办校的过程中，考虑到尽管重庆是陪都，办事很方便，但没有相应的医学教育机构，而成都当时有三所医学院，师资力量强，特别是华西协合大学新建了一所医院，能够提供学生实习的场所。因而当华西协合大学方面主动邀请聂毓禅来华西坝复校并聘请聂毓禅出任该医院护理部主任时，她欣然同意。1943年9月，协和护士学校正式在华西坝复校了，第一批在成都招收了20多名学生。由于协和护士学校的加入，华西协合大学医院的人才力量得到了很大的提高。到了1944年，整个新医院都投入了使用，华西坝上的医科学生都用不着跑到城里去实习了。

Ms. Nie Yuchan graduated from Peking Union Medical College Nursing School in 1927, staying at the school as the deputy head nurse of the hospital ward. From 1929, she went to the Department of Public Health of the University of Toronto and Columbia University Normal College of the USA to study. After returning to China in 1940, Ms. Nie

Yuchan served as the principal of Peking Union Medical College Nursing School, then the only higher nursing school in China. On December 8, 1941 when the Japanese bombed Pearl Harbor, the United States declared war on Japan, and the Peking Union Medical College was forced to close. In the spring of 1943, Ms. Nie Yuchan led the teachers of the nurse school to move interior, hoping to continue to run a nurse school in the rear area. After more than two months hard journey from the occupied Peking, Ms. Nie Yuchan and her group arrived at accompany capital Chongqing. In the process of preparing for the resumption of enrollment of the Peking Union Medical College Nursing School, she considered that although Chongqing was the capital and very convenient to handle affairs, there was no corresponding medical education institution. At that time, Chengdu had three medical universities with strong teaching staff. In particular, West China Union University built a new hospital which could provide a place for students to practice. Therefore, when West China Union University invited Ms. Nie Yuchan and her group to Huaxiba to resume the Peking Union Medical College Nursing School, and hired Ms. Nie Yuchan as the Director of the hospital Nursing Department, she readily agreed. In September, 1943, the Peking Union Nursing School enrolled more than 20 students in Chengdu, and officially resumed its school in Huaxiba. Due to the joining of the Peking Union Nursing School, the talent power of the university hospital had been greatly improved. By 1944, the whole new hospital of West China Union University was put into use, and the medical students in Huaxiba did not need to go to the city for an internship.

经过几年的运行，华西协合大学医院进入了正规发展的轨道，其实力越来越强，即便是抗战胜利以后，1946年聂毓禅带领60多位师生返迁北平，对华西协合大学医院都没有太大影响。1946年年底，重庆大学增设医学院，并把抗战时迁到重庆的中央医院作为重庆大学的教学医院，而中央医院的医务人员都回南京了，留在重庆的医院无法开展正常医疗和教学工作。面对这样的一个局面，重庆方面向华西协合大学求援，华西协合大学校方经过慎重考虑，决定派出以谢锡瑹医生带队的26位医务人员，包括内科、外科、小儿科、产科、妇科和牙科医生，以及放射科、药剂科、护理和人事管理方面的医务人员前往重庆，谢锡瑹任重庆大学医院副院长，兼重庆中央医院代理院长。

After several years of operation, West China Union University Hospital had entered the track of formal development, and its strength was getting stronger and stronger. Even After the victory of the Counter-Japanese War, Ms. Nie Yuchan led more than 60 teachers and students to return to Peking in 1946, which had no impact on the university hospital. At the end of the same year, Chongqing University added a medical school and took the Hospital of Central University School of Medicine moved to Chongqing during the Counter-Japanese War as the teaching hospital of Chongqing University. All the medical staff of the Central Hospital had returned to Nanjing, and the hospitals left in Chongqing was unable to carry out medical and teaching work. Faced with such a situation, Chongqing University asked West China Union University for help. After careful consideration, the university decided to send 26 medical personnel led by Dr. Xie Xishu, including internal medicine, surgery, pediatrics, obstetrics, gynecology and dentists, as well as medical personnel in radiology, pharmacy, nursing and personnel management to Chongqing. Dr. Xie Xishu was appointed Vice Dean of the School of Medicine of Chongqing University and Acting Dean of Chongqing Central Hospital.

1952年全国院系调整以后，随着华西协合大学改名为四川医学院，该医院也随之命名为四川医学院附属医院，人称"川医"。1985年，四川医学院更名为华西医科大学，该医院也被命名为华西医科大学附属第一医院，人称"附一院"。2000年，华西医科大学与四川大学合并，该医院被命名为四川大学华西医院。随着华西医院在老院区附近修建了门诊部大楼以及第一、第二、第三、第四、第五、第六住院大楼等医疗大楼，老院区的老建筑经过重新维修加固作为华西医院的行政楼使用。

After the adjustment of the departments of colleges and universities nationwide in 1952, the West China Union University renamed to Sichuan Medical College, and the university hospital was also named the Affiliated Hospital of Sichuan Medical College, known as "Chuan Yi". In 1985, Sichuan Medical College was renamed West China University of Medical and Sciences. The university hospital was also named the First Affiliated Hospital of West China University of Medical and Sciences, known as "Affiliated First Hospital". In 2000, the University merged with Sichuan University, and the hospital was named West China Hospital of Sichuan University. With the construction and use of outpatient department building, the first, the second, the third, the fourth, the fifth and the sixth inpatient buildings and other medical buildings near the old hospital area by West China Hospital, the old buildings in the old hospital area have been repaired and reinforced and used as the Administrative Building of West China Hospital.

华西坝老建筑的前世今生
The Stories of the Historic Architecture of Huaxiba

拾伍

开男女同校先河

女子大学舍

The Women's College House, the First Coeducation

女子大学舍又名女生院，1924年华西协合大学在中国西部地区开先河，招收了8位女生，实行男女合校，故而有了"男士止步"的女生院。早期女生院修建了两栋小型宿舍供10多位女生住宿，随着越来越多的女生进入大学，女生院也就不断扩建。1932年秋天，在原女生院旁修建了一栋两层的大楼，该楼外形结构为"一字"形，楼的两端和中间均有出入口，屋顶均是中国传统建筑形式歇山顶，屋顶阁楼有中国传统的亮瓦天窗。一楼和二楼都各有连通房间的外走廊，可以容纳70名女生和女教师住宿，称为后院，之前建的宿舍称为前院。1942年，学校又在女生后院修建了一栋两层的大楼，这栋大楼与之前的大楼外形与结构相同并且相向而立，就像一对孪生姐妹。可以说，这两栋楼是女生院的主要建筑。随着大学招收的女生不断增多，为了解决住宿问题，1932年修建的这栋女生学舍的阁楼部分也被改建为寝室，天窗变成了一个长长的老虎窗。

2001年，这两栋大楼被拆除了，修建了容量更大的女生宿舍大楼。

The Women's College House was also known as Girl's Courtyard. In 1924, West China Union University set a precedent in western China by recruiting women and implementing coeducation. So there was a girl's courtyard where "only women could get in". In the early days, two small dormitories were built for more than a dozen girls. As more and more girls entered the university, the girl's courtyard had been expanding. In the fall of 1932, a two-story building was built beside the former girls' dormitories. The shape and structure of the new building was "straight" type with entrances and exits at both ends and in the middle of the building. The roof was the traditional

落成于1932年的女生院大楼，2001年被拆除了。戚亚男摄于1995年。
The girls' dormitory building dedicated in 1932, was demolished in 2001. Photo by Qi Yanan in 1995.

拾伍·开男女同校先河——女子大学舍

女生院大楼屋顶阁楼有中国传统的亮瓦天窗，一楼和二楼都各有连通房间的外走廊。图片来源：四川大学档案馆。

The attic of the girl's courtyard building had traditional Chinese tiled sunroofs. The first and second floors each had outer hallways connecting the rooms. Photo source: Sichuan University Archives.

Chinese architectural form of rest top, and the roof attic had the traditional Chinese bright tile sunroof. Each floor had outer hallways connecting the rooms. It could accommodate 70 girls and female teachers, called the backyard, with the previously built dormitory called the front yard. In 1942, another two-story building was built in the girls' backyard. The building had the same shape and structure as the previous two-story building

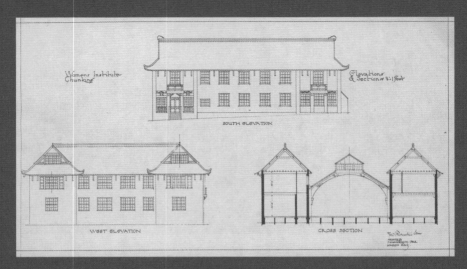

荣杜易设计的女子大学舍立面图和剖面图,从设计图上可以看到该建筑是一对孪生姐妹楼,两栋楼之间有一大厅相连接。遗憾的是最终女子大学舍没有按照设计图修建。赵安祉提供。

Elevations and profiles of the Women's College Houses designed by Mr. Rowntree. It can be seen from the design drawing that the building is a pair of twin sister buildings, with a hall connected between the two buildings. Unfortunately, in the end, the women's college house was not built according to the design. Andrew George provided.

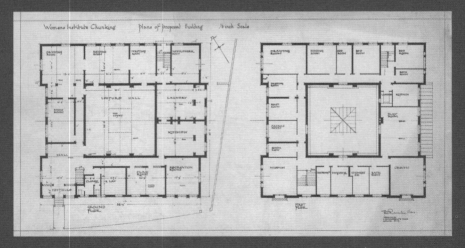

荣杜易设计的女子大学舍平面图,从设计图上可以看到该楼的设计可谓功能齐备,有大厅、客厅、教室、演讲厅、阅览室、写作室、音乐室、娱乐室、卧室、休息室、更衣室、盥洗室、浴室、洗衣房、针线房、厨房、餐厅等。赵安祉提供。

The plan of the Women's College House designed by Mr. Rowntree, it shows that the design of the building is fully functional, including hall, living room, classroom, lecture hall, reading room, writing room, music room, recreation room, bedroom, rest room, dressing room, toilet, bathroom, laundry room, sewing room, kitchen, dining room, etc. Andrew George provided.

and stood opposite, just like a twin sister. It can be said that these two buildings were the main buildings in the girls' courtyard. With the increasing number of female students enrolled in the university, in order to solve the accommodation problem, the attic part of the building built in 1932 had been transformed into dormitories, and the sunroof had been turned into a long roof window.

In 2001, the two buildings were demolished and a larger capacity girls' dormitories were built there.

西部地区第一位女医学博士乐以成
Dr. Yue Yicheng, the First Female Medical Doctor in Western China

1924年9月7日，华西协合大学招收了8位女学生，因此这是西部第一所男女同校的大学。早在华西协合大学筹备之初就酝酿过成立女子学院，设立文科、理科和医科，其目的是让西部的青年妇女也能接受高等教育，然而由于资金问题计划迟迟未能实施。后来学校变通此事，不再计划成立女子学院，而是在华西协合大学里直接招收女生，男女同校，这比单独开办一所女子学院更加进步和开放。乐以成就是西部地区第一批进入高等学府的8位女生之一。

On September 7,1924, West China Union University enrolled eight female students, thus the university was the first coeducational university in western China. As early as at the beginning of the preparation of West China Union University, it was considered to set up a women's college, set up liberal arts, science and medicine disciplines, so that young women in western China could also receive higher education, but it was delayed due to financial problems. Later, the university changed the matter and no longer planned to set up a women's college, but instead enrolled girls directly in West China Union University, which was more progressive and open than opening a women's college alone. Dr. Yue Yicheng is one of the eight first female students to enter higher education institutions in the western region.

乐以成1905年9月26日出生于四川芦山，其父亲乐和洲曾是清朝的武举人。他为人忠厚、正直，辞官不做，归家侍奉父母。

Yue Yicheng was born in Lushan, Sichuan Province, on September 26, 1905. Her father, Yue Hezhou, was once a Wu Ju Ren (military examination graduate) of the Qing Dynasty. He was loyal and upright. He resigned from office and returned home to serve his parents.

尽管乐和洲在当地出资办有一所小学，但在旧的传统思想影响下，乐家的女孩子是不能进学校读书的。看着哥哥甚至弟弟都背着书包去上学，好强的乐以成凭着执着的精神终于让大人们允许她进学校读书。

Although Mr. Yue Hezhou ran a primary school locally, under the traditional thought influence, Yue family's girls could not go to school. Watching her older brothers and even younger brothers carrying schoolbags to school, Yue Yicheng with her Strong character and her persistent

建于20世纪20年代初的女生院,整个院子由4栋建筑组成,用竹篱笆围起。左前是进出女生院的门房,其后是女生的宿舍楼,右边两栋房子是厨房、餐厅、洗漱房等附属用房。图片来源:四川大学档案馆。

The Women's College House was built in the early 1920s. The whole courtyard consisted of four buildings, surrounded by a bamboo fence. On the left front was the gatehouse of the courtyard, and behind was the dormitory building. The two houses on the right were kitchen, dining room, laundry and other ancillary rooms. Photo source: Sichuan University Archives.

spirit, finally let the adults allow her to go to school.

　　乐以成读完了小学,父亲就没有同意她继续读书,让她在家帮助管理家务。无奈乐以成只好留在家里协助父亲管理乐氏大家族,她每天都要做账,家族与外界的书信也都由她来写。就这样乐以成在家干了两年,然而她还是对管理家务事一点都不感兴趣,还是想读书。当时女孩子要想读中学甚至大学就只有到省城去,那里才有女子学校。正好乐以成的一位亲戚在成都华西协合大学工作,从他的口中了解到在成都有女子中学,专门招收适龄的女孩子入校读书。1920年,年仅16岁的乐以成女扮男装悄悄地离家到了成都,进入华英女子中学读书。

　　After Yue Yicheng finished primary school, her father refused to allow her to continue studying, leaving her to help manage the household at home. Yue Yicheng had to stay at home to help her father manage the family. She had to make accounts every day and was responsible the correspondence between the family and the outside world. She worked at home for two years, but she was not at all interested in housekeeping. She still wanted to read. At that time, girls who wanted to go to high school or even university had to go to provincial capital where there were girls' schools. It happened that a relative of Yue Yicheng was working at the West China Union University in Chengdu. She learned from him that there was a girls' middle school in Chengdu, which recruited girls of the right age to study in the school. In 1920, at the age of 16, Yue Yicheng disguised herself as a man, left home quietly to Chengdu, and entered Huaying Girls' Middle School.

　　就在乐以成读中学期间,成都有13位女青年因本地无女子高等学校，只好选择到北平或南京去读书。华西协合大学加快了实施让女青年能接受高等教育的步伐。1924年，华西协合大学决定招收女生，而这一年乐以成也刚好中学毕业，她顺利地考入了华西协合大学文科，成为西部地区第一批进入高等学府的8位女生之一。

During Yue Yicheng's middle school years, 13 young women in Chengdu had no choice but to go to study in Peking or Nanjing because there were no girl's higher education institutions in Chengdu. West China Union University accelerated the pace of implementation to enable young women to enjoy higher education. In 1924, the West China Union University decided to enroll female students. And that year Yue Yicheng also just graduated from middle school. She was successfully admitted to the West China Union University for liberal arts and became one of the first 8 female students to enter higher education institutions in western China.

尽管华西协合大学开先河招收女生，但在那男女授受不亲的社会里，为了避免遭遇社会舆论的非议，学校除了专门修建了女子学舍，还有女舍监负责管理女学生。当时在校读书的男生曹钟梁曾回忆道："当时她们到教室去上课，来去都有一个老妈子送接，避免她们与男生有接触。"所谓"老妈子"就是指负责管理女生的女舍监。

Although the West China Union University took the lead to admit female students, at that time in the society men and women could not be

早期华西协合大学8位女学生在女生院宿舍楼前的门廊上与贝爱理舍监合影。右边戴眼镜站着的女生就是日后西部地区第一位女医学博士乐以成。图片来源：四川大学档案馆。
In the early stage, eight girls of West China Union University took a group photo with supervisor Miss Alice B. Brethorst on the porch of the girls' dormitory. The girl with glasses standing on the right was the first female M.D. in the western region who would graduate from university in the future. Photo source: Sichuan University Archives.

close to each other. In order to avoid criticism from the public opinion, the university, in addition to the special building of women's houses, also hired female wardens in charge of the management. Dr. Cao Zhongliang, a male student at the time, once recalled, "When they went to the classroom, they were sent and picked up by an maidservant to avoid their contact with male students." The so-called "maidservant" refers to the female warden in charge of managing girls.

一年以后，乐以成转到了医科学习，尽管学医科要比其他学科多学几年才能毕业，而且之前没有学过的课程还要补上，但她却认定了要学医当医生。乐以成回忆她小时候在老家看到家族里不少的人生病因无法救治而去世，当医生是她的一个梦想。经过8年的学习，乐以成于1932年毕业了。她不仅获得了华西协合大学医学博士学位，还按照当时学校与美国纽约州立大学的约定，获得了美国纽约州教育委员会颁发的医学博士学位的证书。

A year later, Yue Yicheng transferred to medical studies. Although it took a few years more than other disciplines to finish the courses, and she, on the other hand, had to make up all the courses she had missed, she decided to study medicine to be a doctor. She recalled that when she was a child in her hometown, she saw many of her family members die of incurable diseases. Becoming a doctor was her dream. After 8 years of study, she graduated in 1932. According to the agreement between the West China Union University and the State University of New York at that time, she received both her Medical Doctor degrees from West China Union University and New York State Board of Education.

获得了中外两个学位的乐以成被安排在惜字宫仁济女医院，做了一名妇产科医生。一年以后，乐以成被派往北平协和医院进修一年。在此前后，乐以成遇到了她心中的白马王子——同为学医的谢锡瑹。谢锡瑹是四川璧山县人，之前他在湖南湘雅医学院读书。1928年，湘雅医学院的学生闹学潮，学院被迫关闭，谢锡瑹转学到华西协合大学医科，1931年毕业后被送到北平协和医学院进修放射科和骨科两年多。

Dr. Yue Yicheng, who received two degrees from home and abroad, was arranged to work as an obstetrician and gynecology doctor at the C. M. M. Hospital for Women and Children. A year later, she was sent to Peking Union Hospital for further study. Before and after of this, she fell in love with Dr. Xie Xishu. He was a native of Bishan County, Sichuan Province. He previously studied at Hisiang-Ya Medical College. In 1928, students of Hisiang-Ya Medical College were in a unrest and the college was forced to close down. He transferred to West China Union University Medical School, graduated in 1931 and was sent to Peking Union Hospital to study radiology and orthopedics for more than two years.

乐以成与谢锡瑹从北平协和医学院进修回到成都后，分别在惜字宫仁济女医院和四圣祠男医院工作。1940年因女医院失火被毁，女医院就与男医院合并，乐以成遂与丈夫在同一家医院工作。作为大学的教学医院，对人才的培养是很重视的，各科都要选派医生到国外进修,学习最先进的医疗技术,乐以成先后被派往加拿大多伦多大学、美国加利福尼亚大学和英国伦敦大学皇家医学院进修。乐以成在伦敦进修期间正值中华人民共和国成立,她收到了丈夫谢锡瑺叫她回国的信。同时,她也得到了英国移民局的邀请留在英国,并且让她写信请她丈夫带儿女也到英国定居。乐以成经过再三思考并与丈夫商量,最终于1950年年底回到了成都。

After Dr. Yue Yicheng and Dr. Xie Xishu returned to Chengdu from Peking Union Hospital, they worked in the C. M. M. Hospital for Women and Children and the General Hospital respectively. In 1940, when the Hospital for Women and Children was destroyed by a fire, the Hospital for Women and Children was merged with the General Hospital, and Dr. Yue Yicheng worked in the same hospital with her husband. As a teaching hospital of the university, it attached great importance to the cultivation of talents. All departments should send doctors to study abroad and learn the most advanced medical technology. Dr. Yue Yicheng had been sent to the University of Toronto, the University of California and the Royal School of Medicine of the University of London for further study. During she studying in London, just when the People's Republic of China was founded, she received a letter from her husband, asking her to return China, and she also received an invitation from the British Immigration Office to stay in the UK, and asked her to write a letter asking her husband to take their children to settle in the UK. After careful thinking and consultation with her husband, she finally returned to Chengdu at the end of 1950.

乐以成从医以来为许许多多的产妇接生过,有人把她与北平协和医院我国妇产科泰斗林巧稚相提并论,人称"北有林巧稚,南有乐以成"。有一天,乐以成在成都春熙路上遇到了一位50多岁的男士,该男士上前不容分说地握住乐以成的手说:"乐婆婆,你还认得我吗?幸亏你50年前为我接生……""你是男的,我咋会给你接生呢?"乐以成笑着回答道。经过交谈才得知,原来当年这位男士出生时其母亲难产,多亏了乐以成才保住了母子的平安。

Dr. Yue Yicheng had delivered many babies in her medical career. Someone compared her to Dr. Lin Qiaozhi of Peking Union Hospital. Dr. Lin was the authority of obstetrics and gynecology in China. People said "There are Lin Qiaozhi in the north and Yue Yicheng in the South". One day, Dr. Yue Yicheng met a man in his 50s on Chunxi Road in Chengdu. The man came forward to hold her hand and said, "Grandma Yue, do you still recognize me? Fortunately, you delivered me 50 years ago…" "You are a man. How can I deliver you?" she replied with a smile. After talking, she learned that when the man was born, his mother had difficulty labour. Thanks to Dr. Yue's success, she saved the mother and son.

20世纪90年代初,一位英国剑桥大学的教授来成都讲学,他专门带上夫人和女儿登门向乐以成致谢,感谢乐以成在40多年前救了他夫人和女儿的命。原来当年这个英国教授在成都教书时,夫人临产前先兆子痫、妊娠高血压、水肿送到四圣祠医院时情况已十分危险,乐以成果断地为产妇做了剖宫产手术,最终母子平安。

后来教授一家人回到英国后，其夫人又顺产过两胎，当时接生的英国大夫对乐以成之前做的剖宫产手术很是称赞，因为产妇在剖宫产手术后再怀孕生产很难自然正常分娩。从这位教授夫人的案例中可以看出乐以成的医术是非常高的。

In the early 1990s, a professor from Cambridge University of England came to Chengdu to lecture. He brought his wife and daughter to thank Dr. Yue Yicheng for having saved his wife and daughter more than 40 years ago. It turned out that when the British professor was teaching in Chengdu, his wife suffered from preeclampsia, pregnancy hypertension, edema and convulsions before delivery. Her situation was very dangerous when she was sent to the C. M. M. Hospital for Women and Children. Dr. Yue Yicheng decisively performed a caesarean section for the puerpera, and both the mother and baby were safe. Later, the professor's family returned to the UK, and his wife gave birth to two children naturally. The British doctor praised the caesarean section that Dr. Yue Yicheng did for the professor's wife, because it was difficult to give birth naturally after a caesarean section. From this case we can see that Yue Yicheng's medical skill is very high.

乐以成从医60多年来给无数的家庭带来了欢乐，85岁高龄还在看门诊。2001年4月28日，96岁的乐以成离开了我们。

Dr. Yue Yicheng had been a doctor for more than 60 years bringing joys to countless families and still worked in outpatients clinic at the age of 85. On April 28, 2001, she left us at age 96.

然而不为外人所知的是，乐以成的女儿谢蜀祥继承了她母亲的事业，也是一位优秀的妇产科医生。乐以成对她的女儿既有慈母的关爱，更有老师的严格。谢蜀祥说："我跟我妈妈之间是双重关系，不仅是母女关系，而且我是她的徒孙——她学生的学生。我与我妈一起共事了近30年，我对她是备感思念的，我很敬佩她，我也很害怕她，我也很热爱她。"

However, unknown to outsiders, her daughter Xie Shuxiang is also an excellent obstetrician and gynecologist. To her daughter, Dr. Yue is not only a loving mother, but also a strict teacher. Doctor Xie said, "I have a double relationship with my mother, not only as a mother and daughter, but also I am a student of her. I have worked with my mother for nearly 30 years, and I miss her deeply. I admire her, I am also very afraid of her, and I love her very much."

抗战时期五大学国际救护队队员黄孝逴烈士
Huang Xiaochuo Martyr, an International Rescue Team Member

1939年，国民政府教育部令华西协合大学在女生院里修建了一座纪念碑，以纪念被日军炸死的华西坝五大学国际救护队队员女生黄孝逴。

In 1939, the Ministry of Education of the National Government ordered West China Union University to build a monument in the girls' courtyard to commemorate Ms. Huang Xiaochuo, a member

华西协合大学理学院制药系学生黄孝逴。黄孝述提供。
Huang Xiaochuo, a student of the Department of Pharmacy, School of Science, West China Union University. Huang Xiaoshu Provided.

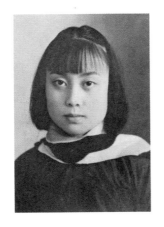

of the International Rescue Team of Huaxiba Associated Universities in Chengtu, who was killed by the Japanese army.

黄孝逴是四川永川人，父亲黄默涵早年留学日本。黄孝逴1916年出生于上海，1935年随全家迁到成都，1937年考入华西协合大学理学院制药系。时值抗日战争时期，黄孝逴在学校里参加了"华西学生救亡剧团"，利用节假日到校外宣传抗日。她还与同学们一起为前线将士缝制冬装，参加义卖献金活动。

Ms. Huang Xiaochuo was a native of Yongchuan, Sichuan Province. Her father studied in Japan in his early years. Ms. Huang Xiaochuo was born in Shanghai in 1916, moved to Chengdu with her family in 1935. Ms. Huang Xiaochuo was admitted to the Pharmaceutical Department of School of Science, West China Union University in 1937. During the Counter-Japanese War, she participated in the "Huaxi Student National Salvation Troupe" at school, and used holidays to publicize counter-Japanese outside the university. She also worked with her classmates to sew winter clothes for soldiers at the front and donated money through charity sales.

自从1938年11月8日18架敌机首次空袭成都以来，全市就加强了应对日军轰炸的方案。第二年开春，在华西坝五大学战时服务团的倡导下决定成立五大学国际救护队，很快华西坝上五大学就有300多位同学报名参加。中央大学医学院解剖学教授张查理、神经科教授程玉麟和教师陆振山，齐鲁大学医学院外科医生刘荣耀等老师报了名。高子厚、启真道和吉士道等外籍教授也参加了救护队。大家公推张查理教授为大队长，华西协合大学医牙学院学生吴廷椿、齐鲁大学医科学生刘春林和中央大学医学院的一位同学分别担任中队长，每个中队下又分为3个小队。华西协合大学医牙学院院长启真道等外籍教授还为救护队提供担架、药品等急救用品，安排队员学习、训练怎样急救。

Since the first air strike of 18 enemy aircraft on Chengdu on November 8, 1938, the city had strengthened its plan to deal with Japanese bombing. In the spring of the next year, under the initiative of the wartime service group of Huaxiba Associated Universites in chengtu, the university decided to set up the International Rescue Team of Huaxiba Associated Universities in Chengtu. Soon, more than 300 students from Huaxiba Associated Universities in Chengtu joined. Zhang Charlie, Professor of Anatomy at the Medical School of Central University, Lu Zhenshan, Cheng Yulin, Professor of Neurology, and Liu Guangrong, surgeon at the Medical School of Cheeloo University, and other teachers signed up. Foreign professors such as Dr. W. Crawford, Dr. L. G. Kilborn and Dr. H. J. Mullett also joined the Rescue Team. Professor Zhang Charlie was nominated as the team leader. Wu Tingchun, a student of the Medical and Dental College of West China Union University, Liu Chunlin, a student of medical school of Cheeloo University, and a student of Central University Medical School were appointed as the squadron leader respectively, and each squadron was divided into three teams. The Dean of the Medical and Dentistry College of West China Union University L. G. Kilborn and other foreign professors, also provided stretchers, drugs and other first-aid supplies for the Rescue Team, and arranged the team members to learn and train how to do first aid.

黄孝逴与华西协合大学医牙学院的牙科学生陶海鹏分在一个小队里，在救护队里参加急救训练时陶海鹏总是当黄孝逴练习包扎被炸伤的伤员，这样他们俩就相互认识了。课余时间陶海鹏时常到黄孝逴家里去玩耍，一起去华西坝的青春岛划船、游泳。到后来陶海鹏每周二、周五晚饭后都去女生院找黄孝逴，他们一起交流校园里发生的事情，有时也说说时局。谈到日本人轰炸重庆、成都时，陶海鹏对黄孝逴说如果敌机来成都轰炸，她就到他住的广益学舍地下室躲避。讲这话时，陶海鹏忘了他们俩是救护队队员，敌机来轰炸时，他们必须要去抢救伤员。

Ms. Huang Xiaochuo and Mr. Tao Haipeng, a dental student at the Medical and Dental College of West China Union University, were in a team. When attending first aid training in the Rescue Team, Mr. Tao Haipeng always was the wounded when Ms. Huang Xiaochuo practiced dressing the wounded. So the two knew each other. In his spare time, Mr. Tao Haipeng often went to Ms. Huang Xiaochuo's home to play and went to boat and swim on the Youth Island of Huaxiba together. Later, Mr. Tao Haipeng went to the reception room of girls' courtyard to date Ms. Huang Xiaochuo every Tuesday and Friday after dinner. They talked about what was happening on campus and sometimes about the current political situation. When talking about the Japanese bombing of Chongqing and Chengdu, Mr. Tao Haipeng told Ms. Huang Xiaochuo that if the enemy aircraft came to Chengdu to bomb, she could come to hide in the basement of the Friends' College Building where he lived. When saying this, Mr. Tao Haipeng ignored that they were two members of the Rescue Team, and when the enemy aircraft came to bomb, they should go to rescue the wounded.

然而黄孝逴与陶海鹏两人的交往很快就因日本轰炸机的轰炸永远地结束了。1939年6月11日星期天，傍晚7点过，27架日本轰炸机轰炸成都，市区响起了空袭警报，陶海鹏很快就赶到了救护队。当时黄孝逴正与华西协合大学周芷芳、齐鲁大学崔之华在学校附近一家餐馆吃饭，听到警报声她们立刻放下碗筷，离开餐馆准备到学校救护队集合地点集结。这时日本轰炸机已经投下了炸弹，其中有4枚炸弹落在了华西坝上，就在黄孝逴她们快要进校门时，一枚流弹击中了黄孝逴的后脑，她当场身亡。而此时陶海鹏正与救护队队员把被炸伤员抬到学校事务所做急救处理。郭友文队长告诉他黄孝逴中弹去世了，他还不相信，直到队长叫陶海鹏去看黄孝逴的遗体他才确信。陶海鹏见过黄孝逴的遗体后，立刻赶到黄孝逴的家里把这个噩耗告诉了她的家人。黄孝逴遇难后，陶海鹏心情很不安，他放弃了要学6年的医牙学院而转入了只学4年的理学院化学系，期望早日毕业。

However, the contact between Ms. Huang Xiaochuo and Mr. Tao Haipeng soon ended forever due to the bombing of Japanese bombers. On Sunday, June 11,1939, after 7 p. m., 27 Japanese bombers bombed Chengdu, and an air raid alarm sounded in the city. Mr. Tao Haipeng soon arrived at the Rescue Team. At that time, Ms. Huang Xiaochuo was having dinner with Ms. Zhou Zhifang, a classmate of West China Union University, and Ms. Cui Zhihua, a classmate of Cheeloo University in a restaurant near the university. When they heard the alarm, they immediately put down their bowls and chopsticks, left the restaurant and returned to the university to assemble at the assembly site of the Rescue Team. By this time, the Japanese bombers had dropped bombs, four bombs fell on the Huaxiba, when Ms. Huang Xiaochuo and her classmates were about to enter the university gate, a stray bomb hit Ms. Huang Xiaochuo in the back of the head, she died on the spot. At this time, Mr. Tao Haipeng and the rescue team members carried the injureds to the university office for emergency treatment. The team leader Guo Youwen told him that Ms. Huang Xiaochuo was hit and dead. He did not believe it until the team leader asked Mr. Tao Haipeng to see Ms. Huang Xiaochuo's body. After seeing Huang Xiaochuo's body, Mr. Tao Haipeng immediately rushed to Ms. Huang Xiaochuo's home to tell the bad news to her family. Since Ms. Huang Xiaochuo died, Mr. Tao Haipeng was very upset. He gave up the Medical and Dental College which needs six years to study and transferred to the Chemistry Department of the School of Science which only needs four years, hoping to graduate as soon as possible.

事后《新新新闻》报刊发了题为《前日寇机袭蓉华西大学中弹数枚该校电英美各国向倭提出严重抗议》的文章，文中写道："黄孝逴参加国际救护队于敌机来袭时出发冒险工作不顾己身之艰危，致遭不幸，中弹身死。"消息传到重庆，国民政府教育部得知后通令嘉奖黄孝逴，并特发国币五百元，由学校立碑纪念。

Later, the *New New Newspaper* issued an article titled *The Former Japanese Bombers Attacked Chengdu City, and the West China University Was Hit by Several Bombs. The University Telegraphed British, America and other Countries Filed Serious Protests against the Japanese Pirates*, which wrote, "Huang Xiaochuo joined the International Rescue Team and set out to work at risk when the enemy plane attacked regardless her own difficulties and dangers, she was unfortunately hit and killed." After the news came to Chongqing, the National Ministry of Education order commendatory to reward Huang Xiaochuo, and specially gave national currency 500 yuan to the university to set a monument of Huang Xiaochuo.

1940年，黄孝逴遇难一周年之际，在华西坝女生院修建了一座日晷式样的纪念碑，该碑有半人高，六边形，有三层台阶，碑的六面刻有碑文。随着时代的变迁，建在女生院的黄孝逴纪念碑在20世纪50年代就被拆除了。

In 1940, on the first anniversary of Ms. Huang Xiaoyun's death, a sundial shaped monument was built in Huaxiba girls' courtyard. The monument was half man high, hexagonal, and has three steps. Inscriptions were engraved on the six sides of the monument. As the times change, the monument to Huang Xiao, built in the girls' courtyard was demolished in the 1950s.

女生院的五月花柱节
The Maypole Festival at the Girls' Courtyard

自从1924年华西协合大学开中国西部地区先河招收女生而修建了女生院以来，因为该院封闭式管理，从不对外开放，有人称其为"禁宫"。起初女生离开女生院到教室里去上课都有女教士陪同前往，后来女生上课没有人陪同，她们可以自由进出女生院，但男士是不能进入的。

Since 1924, West China Union University opened the first college for girls in western China, for it was closed to public, some people called it a "Forbidden Palace". At first, the girls left the girls' courtyard to the classroom accompanied by female missionaries. Later, the girls were not accompanied by female missionaries to classroom, so they could freely enter and out of the girls' courtyard, but the men could not enter the girls' courtyard.

随着社会文明的进步，到了20世纪40年代，有更多的女孩子进入学校读书，毕业以后成为职业女性，女生院的大门也慢慢地向社会开放了。通常每年5月第一周的星期六下午，女生院要举办一个源于国外的"五月花柱节"，在这一天，女生院的女生在院里举办舞蹈表演，允许男士进入参观，但要凭券进入。两栋相向而对的女生宿舍楼的外走廊就是看台，中间的草坪就是一个天然的露天大舞台，一架钢琴被抬到室外做伴奏。

With the opening of social civilization, in the 1940s, more girls entered the university and became professional women after graduation. The door of the girls' courtyard was also slowly opened to the society. Usually, on the Saturday afternoon of the first week of May every year, the girls' courtyard would hold "The Maypole Festival" coming from abroad. On this day, the girls of the girls' courtyard would hold a dance performance in the courtyard, allowing men to enter and visit, but with tickets. The outer hallways of the two girls' dormitory buildings standing opposite each other were the grandstand. The middle lawn was a natural open-air stage, and a piano was carried outdoors for accompaniment.

据当地新闻报道，1947年5月的花柱节里女生们表演了兵舞、采豆舞、快乐舞、黑人舞、俄国土舞、干戈舞、英国土风舞、轻快舞、复兴舞、宫廷舞、海盗舞、莱士舞和五月花柱舞等。最精彩的表演是五月花柱舞。女生们围着一根插在地上的高高的竹竿，竹竿的顶端拴有多条黄绿两色飘带，女孩子们手持彩带的另一端

女生院每年的花柱节最精彩的表演是五月花柱舞,女生们围着一根插在地上的高高的竹竿跳舞,竹竿的顶端系有黄绿两色飘带,女孩子们手持彩带的另一端围着竹竿翩翩起舞。图片来源:耶鲁大学提供。

The most wonderful performance was the Maypole dance in the Maypole Festival. The girls surrounded a tall bamboo pole inserted on the ground. The top of the bamboo pole was tied with a number of yellow and green long ribbons. The girls held the other end of the ribbon and danced around the bamboo pole. Photo source: Yale University.

围着竹竿翩翩起舞。舞蹈表演后通常还要演一出舞剧,五月花柱节才算结束。1949年的五月花柱节有近2 000人前去观看,大家兴致勃勃地看完了舞蹈表演后,正准备观看女生们精心准备的舞剧《罗密欧与朱丽叶》时,天公不作美,突然雷声大起,大雨大风随之而来,演出只好停止。舞剧改在4天后演出,到时观众持原有的入场券到女生院观看。

According to local news reports, at the Maypole Festival in 1947, girls performed soldier dance, bean picking dance, happy dance, black dance, Russian national dance, Gan Ge dance, British folk dance, light dance, revival dance, court dance, pirate dance, Laishi dance and Maypole dance. The most wonderful performance was the Maypole dance. The girls surrounded a tall bamboo pole inserted on the ground. The top of the bamboo pole was tied with a number of yellow and green long ribbon. The girls held the other end of the ribbon and danced around the bamboo pole. After the dance performance, there was usually a dance drama, then the Maypole Festival was over. In 1949, Nearly 2,000 people went to watch the Maypole Festival. After watching the dance performance with great interest, they were preparing to watch the dance drama *Romeo and Juliet* carefully prepared by the girls. Suddenly loud thunder burst, and heavy rain and wind followed, and the performance had to stop. The dance drama performed in four days later. At that time, the audience would hold the original admission ticket to the girls' courtyard to watch it.

女生院让在华西坝上读书的女孩子们留下了不少美好的回忆，时常有满头白发的女士或几位一起，或在家人的陪同下来到女生院寻找往日的身影。最让笔者记忆尤深的是1993年秋，国画大师徐悲鸿的夫人廖静文来成都举办徐悲鸿主题画展，画展期间廖静文女士特意来到华西坝寻找半个世纪前的美好记忆。1943年廖静文考入了成都金陵女子文理学院化学系，就住在女生院里。那一年徐悲鸿在青城山写生，时常来这里与廖静文约会。故地重游勾起了廖静文女士对过去的甜蜜回忆，她站在女生院宿舍楼下，脸上洋溢着幸福的光彩说："当年我就住这里，悲鸿就经常在楼下等我。"

The girls' courtyard had left many beautiful memories for the girls studying in Huaxiba. Often, there are white haired women or a few together, or accompanied by their family, come to the girls' courtyard to look for their past figure. What the author remembered most was that in the autumn of 1993, Ms. Liao Jingwen, the wife of master of traditional Chinese painting Xu Beihong, came to Chengdu with paintings to hold Xu Beihong Theme Painting Exhibition. During the exhibition, Ms. Liao Jingwen specially came to Huaxiba to look for the beautiful memory of half a century ago. In 1943, Ms. Liao Jingwen was admitted to the Chemistry Department of Chengdu Ginling College and lived in the girls' courtyard. That year, Mr. Xu Beihong was sketching in Qingcheng Mountain and often came to the girls' courtyard to meet Ms. Liao Jingwen. The visit brought back sweet memories of the past for Ms. Liao Jingwen. Standing downstairs of the dormitory of the girls' courtyard, her face was filled with the glow of happiness and said, "When I lived here, Beihong often waited for me downstairs."

华西坝老建筑的前世今生
The Stories of the Historic Architecture of Huaxiba

拾陆

华西坝上的教职员住宅

小洋楼

The Faculty and Staff Residence,
Small Foreign Style Houses in Huaxiba

华西坝老建筑的前世今生（汉英对照）
The Stories of the Historic Architecture of Huaxiba (Chinese-English Bilingual Edition)

当年英国建筑设计师荣杜易在设计华西协合大学校园时，就在设计图上规划了在校园西南边呈弧形排列修建十几栋小巧的一楼一底住宅供教师居住，这些住宅与教学楼一样也具有西方建筑元素，因而被市民称为"小洋楼"。

When the British Architectural designer Fred Rowntree designed the campus of West China Union University, he planned to build more than dozen small two-story residential buildings for teachers to live in. The houses arranged in an arc in the southwest of the campus. Like the teaching buildings, these houses also had western architectural elements, so they were called "small foreign style houses" by the citizens.

华西坝上这些中西合璧的建筑影响了20世纪20至40年代成都的民居建设，甚至在川内其他县市也有模仿华西坝上的一些建筑而建造

20世纪40年代航拍的华西坝鸟瞰图，照片左边为懋德堂图书馆、怀德堂办公楼、教育学院，中间为合德堂和嘉德堂生物楼、化学楼，右边为钟楼，右下角呈弧形排开的有8栋小洋楼，靠近钟楼的为1号楼，就是校长居。图片来源：四川大学华西医学中心提供，芳卫廉的女儿赠。

Aerial view picture of Huaxiba in 1940s, Library Building, Administration Building and College of Education were on the left, Hart College, Biological Building and Chemical Building were in the middle, the Clock Tower was on the right. There were 8 small foreign style buildings arranged in an arc in the lower right corner, and building No. 1 is the President Residence near the clock tower. Photo source: The West China Medical Center, Sichuan University, a gift from William P. Fenn's daughter (Mary Francis Hazeltine).

上图：这张照片展示了围绕校园的4座小洋楼，远处左上角的塔楼就是合德堂，中间露出屋顶的是生物楼，此时钟楼还未建成，这张照片的拍摄年代应该是在1924—1926年。柯马凯提供。

The photo shows four small houses around the campus, the tower of Hart College in the upper left corner of the far distance and the roof of the Biological Building in the middle of the far distance. At this time, the clock tower had not yet been built. So, this photo was supposed to be taken between 1924 and 1926. Provided by Michael Crook.

下图：这张照片里有4座小洋楼掩映在树林中，远处是高大的钟楼。这张照片与上面那一张照片是在同一角度拍摄的，只是镜头稍微向右移动了一点，因而看不见合德堂上的塔楼，生物楼也被高大的树林掩盖了屋顶，而1926年建的钟楼却高耸在远处。从植物的茂密程度来看，这张照片估计拍摄于20世纪40年代。柯马凯提供。

In this photo, there are four small houses hidden in the woods, and the tall Huaxi Clock Tower is in the distance. The photo was taken at the same angle as the one above, with the camera moving slightly to the right so that the tower of Hart College was not visible. The Biological Building was also covered by the tall woods. But the clock tower built in 1926 was towering in the distance. Judging from the density of plants, this photo was estimated to have been taken in the 1940s. Provided by Michael Crook.

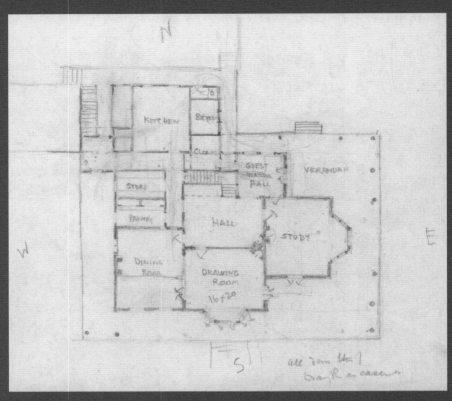

上图：荣杜易手绘的小洋楼平面图草稿。赵安祝提供。
A hand-painted draft of the floor plan of the Small Foreign Style Houses by Mr. Rowntree. Andrew George provided.

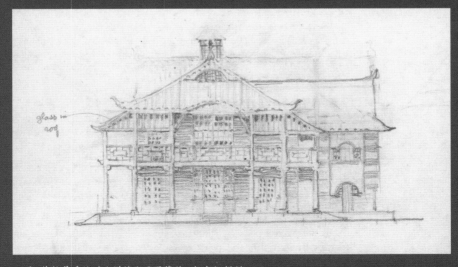

下图：荣杜易手绘的小洋楼立面图草稿。赵安祝提供。
A hand-painted draft of the Small Foreign Style House facade by Mr. Rowntree. Andrew George provided.

的楼房。当年在华西坝上修建的近50栋小洋楼到现在已经所剩无几,好在仅存的3栋被挂牌保护起来了。

These buildings of combination of Chinese and Western style in Huaxiba, influenced the residential construction in Chengdu from 1920s to 1940s. Even other counties and cities in Sichuan had buildings built by imitating some buildings in Huaxiba. At that time, there were nearly 50 houses in Huaxiba, but now only few left. Fortunately, the only three remaining ones have been listed for protection.

华西坝上最早的小洋楼校长居
The Earliest Small Foreign Style House in Huaxiba

1910年华西协合大学建校之初,学校除了修建了一批临时的教室,还为外籍教师修建了一些临时的简易住宅。直到1912年荣杜易获得了大学校园建筑设计赛竞标后,校舍建筑就全面展开修建了。华西坝上最早修建的小洋楼是在1916—1917年,这是据赵安祝提供的一张正在修建的小洋楼照片确定的,这张照片里显示的小洋楼已经快要完工,其背面写的时间是1916年,以及居住者的名字为吴哲夫。之后这幢小洋楼编号是1号楼,一直都由学校校长居住,称之为"校长居"。

In 1910, when West China Union University was founded, in addition to building a number of temporary classrooms, it also built some temporary houses for foreign teachers. It was not until 1912, when Mr. Rowntree won the competition to design the university campus building, that the construction of the university building began in full force. Based on a photograph of a small building in progress provided by Dr. Andrew George, the earliest small foreign style house built in Huaxiba dated from 1916 to 1917. The photo showed the small building about to be completed 1916 and the name of the resident E. W. Wallace were written on the back. Later, this small foreign style house was numbered No.1 small house, which had always been lived by the university president, known as the President Residence.

吴哲夫1904年毕业于多伦多大学维多利亚学院,他对中国很感兴趣,1903年他写了《筚路蓝缕——西医入川》一书。1906年他来到成都,积极参与中国西部地区各教派的教育联合。华西协合大学成立后,吴哲夫任教育科主任,之后担任大学副校长,同时他还出任华西教育会的总干事。1929年吴哲夫从上海返回加拿大多伦多。

Mr. Wallace graduated from Victoria College at the University of Toronto in 1904 and was very interested in China. In 1903, he wrote a book titled *THE HEART SZ-CHUAN*. In 1906, he came to Chengdu. He actively participated in the educational union of various denominations in western China. After the founding of West China Union University, Mr. Wallace served as Director of the Education Department and later as Vice President of the University. At the same time, he

also served as General Director of the Educational Union of West China. In 1929, Mr. Wallace returned to Canada from Shanghai and became the Dean of Victoria College at the University of Toronto, Canada.

自从吴哲夫离开华西坝后，这栋小洋楼就由校长毕启入住。毕启入住小洋楼后一刻都没有放松学校的建设和发展工作，他时常与学校同仁在校园里稻田的田埂上散步，畅想着大学未来的校园建设。

Since Mr. Wallace left Huaxiba, the university President Beech moved in the President Residence. After moved into the small house, Dr. Beech never relaxed about the construction and development of the university. He often walked with his colleagues on the ridge of the rice field on the campus, imagining the future campus construction of the university.

在毕启担任校长和校务长的30多年里，他经常穿越长江、横渡大西洋和太平洋，往返于欧洲和美国，为大学筹集资金。到1940年毕启73岁退休回美国时，大学已经初具规模，校园里修建了20栋大楼、50栋教职员宿舍，这都是毕启募集资金而建造的，此外还包括正在修建的大学医院和化学大楼。

"校长居"的大门为典型的中国传统建筑风格歇山式屋顶，两旁建有耳房，大门左右的墙面有中国建筑特有的漏花窗形式图案。赵安祝提供。
The gate of the President Residence is a typical gable and hip roof of traditional Chinese architectural style, with ear rooms on both sides of the gate, and the wall with leaky flower window patterns unique to Chinese architecture. Provided by Andrew George.

校长居的大门拆去了左边门墙，把门扩大了，这样一来看上去不对称，很别扭，不过可以让小汽车有足够的空间进出。戚亚男摄于2011年。

The left part of the gate wall of the President Residence was removed and the gate was expanded. In this way, it looks asymmetric and awkward, but it can allow cars to have enough access. Photo by Qi Yanan in 2011.

During more than 30 years as president and university governor, Dr. Beech frequently went through Yangtze River, and crossed Atlantic Ocean and Pacific Ocean many times, to and fro between Europe and the USA, to raised funds for the university. By 1940, when he retired at the age of 73 and returned to the United States, the university had been built in great scale. There were 20 teaching buildings and about 50 houses for staff in campus. All these was built with money raised by Mr. Beech, including the university hospital and the chemical building under construction.

毕启退休回美国后，张凌高校长就入住了"校长居"。张凌高是华西协合大学首任华人校长。

After Dr. Beech retired and went back to the United States, the president Zhang Linggao moved to the President Residence. He was the first Chinese president of the West China Union University.

20世纪50年代以后，在"校长居"居住过的有四川医学院副院长、党委书记孙毅华，教务长蒋旨昂，副院长顾锷。20世纪80年代以后"校长居"就成为学校的招待所。

After the 1950s, Mr. Sun Yihua, vice president and secretary of the Communist Party Committee of Sichuan Medical College, provost Jiang Zhiang, and Mr. Gu E, vice president of Sichuan Medical College had lived in the President Residence. Since the 1980s, the President Residence has been used as a hostel of the university.

2015年3月，毕启的后代寻根访问团一行12人专门来华西坝寻根，他们把爷爷毕启曾经使用过的"校长居"的钥匙送给了四川大学华西医学中心。毕启的孙子在爷爷曾经办公和居住的小洋楼前回忆起当年的往事说道："当我小的时候，在家中的老照片上就看到过华西校区的这些很特别的建筑。""今天实实在在看到了，觉得很亲切，有一种回家的感觉。""这里是爷爷曾经的住所，对于这里这把钥匙很有象征意义，它应该属于这里。"

In March 2015, 12 members of Dr. Beech's descendants came to Huaxiba to search for their roots. They gave the key of the President Residence once used by Dr. Beech to West China Medical Center of

保留完整的小洋楼，只是在左边为了适应招待所的功能而添加了附属的用房。戚亚男摄于2011年。

Keep the complete small western-style house, but added auxiliary rooms on the left to adapt to the functions of the guest house. Photo by Qi Yanan in 2011.

Sichuan University. In front of the small house where his grandfather once worked and lived, Dr. Beech's grandson recalled the past and said, "When I was young, I saw these special buildings on West China campus in the old photos at home." "I really saw it today, and I feel very warm. I feel at home." "This is where grandpa used to live. This key here is very symbolic. It should belong here."

2017年7月,"校长居"被成都市确定为历史建筑并挂牌保护。

In July, 2017, the President Residence was identified as a historical building and listed for protection by Chengdu Municipal People's Government.

华西坝上的苏木匠
Carpenter Su of Huaxiba

2016年12月,成都市人民政府在华西坝"校长居"旁的一栋小洋楼前立了"成都市历史建筑苏继贤旧居"的牌匾,牌匾上写道:"建于20世纪20年代。原华西协合大学加拿大籍建筑工程师苏继贤的住宅。原华西协合大学教育系主任饶和美与原华西大学校长、动物学家、教育学家、我国两栖爬行动物学的主要奠基人之一刘承钊曾在此居住。中西合璧建筑风格。"

In December, 2016, the Chengdu Municipal People's Government set up a sign of "Chengdu Heritage Architecture, W. G. Small's Old Residence" by a small house next to the President Residence of Huaxiba. The sign says, "Built in the 1920s, this house was a blend of Chinese and Western styles. It first served as the residence of Walter Small, a Canadian architect working at the former West China Union University, and was later the home of Homer G. Brown, Head of the Education Department of that University. Liu Chengzhao, President of the former West China University, also a zoologist, educator and one of the founders of Herpetology in China, once lived here too."

苏继贤旧居是华西坝2号小洋楼。2号小洋楼的第一位居住者是英国人苏道璞。1913年,苏道璞被英国公谊会派到成都华西协合大学担任化学教授,1928年出任大学副校长一职。1930年5月30日晚,苏道璞在校内骑自行车经过合德堂时,遭遇抢劫而受伤,因抢救无效,于6月1日死亡。为了纪念遇害的苏道璞副校长,1941年秋落成的四所大学联合修建的化学楼被命名为苏道璞纪念堂。

Walter. G. Small's Old Residence was No. 2 small house. The first occupant of the No. 2 small house was Dr. C. M. Stubbs. In 1913, Dr. Stubbs was sent to Chengdu West China Union University as a professor of chemistry by the British Friends Foreign Mission Association. In 1928, he served as vice president of the university. On the evening of May 30, 1930, Dr. Stubbs was injured by robber while riding a bicycle in university. He was not rescued and died on June 1. A chemical building jointly built by four universities during the Counter-Japanese War, which was completed in the autumn of 1941, was named C. M. Stubbs Memorial Hall in memory of the murdered vice president Stubbs.

之后，2号楼入住的是教育系主任饶和美一家人。饶和美1920—1939年在华西坝上任教，其夫人饶珍芳1919年接手管理成都私立弟维小学，为了让学龄前儿童有专门的机构管理，让其母亲们能进入社会工作，她特意在弟维小学新设了一所幼稚园（现成都市第十一幼儿园）。饶和美和饶珍芳夫妇1915年在成都出生的女儿伊莎白就是在这栋小洋楼里长大的。1933年，18岁的伊莎白从华西坝的加拿大学校完成了她的高中学业后回到加拿大，就读于多伦多大学维多利亚学院。1938年获得硕士学位后，伊莎白回到四川从事人类学研究。1948年，应中共中央外事组副组长王炳南邀请，伊莎白加入了刚成立的外事干部培训学校从事英语教学工作。1949年中华人民共和国成立了，外事干部培训学校迁到北京，随后改名为外国语学校（即现在的北京外国语大学）。自此以后伊莎白就留在了中国，是新中国英语教学的拓荒人。2019年9月17日，由国家主席习近平签署主席令，授予北京外国语大学加拿大友人、老专家伊莎白中国国家对外最高荣誉勋章——中华人民共和国友谊勋章。

落成后的2号小洋楼，整座楼的东、南、北三面都是连通的阴台，照片右边为合德堂，摄于1921年。赵安祝提供。

After the completion of the No. 2 small house. The east, south and north of the whole building are connected by the long corridor. On the right of the photo is Hart College. Taken in 1921. Provided by Dr. Andrew George.

After that, the Education Department Director Homer. G. Brown family moved in to the No. 2 small house. Mr. Brown taught at Huaxiba from 1920 to 1939. His wife Muriel. H. Brown took over the management of Chengdu Private Dewey Primary School in 1919. In order to let preschoolers had special institutions to manage and let their mothers to work, she specially set up a new kindergarten in Dewey Primary School (now Chengdu No.11 Kindergarten). The couple's daughter Isabel born in Chengdu in 1915, grew up in this small house. In 1933, 18-year-old Isabel returned to Canada after completing her high school studies from the Canadian School in Huaxiba to study at Victoria College of the University of Toronto. After obtaining her master's degree in 1938, Isabel went to Sichuan to engage in anthropological research. In 1948, invited by Mr. Wang Bingnan, deputy leader of the Foreign Affairs Group of the CPC Central Committee, Ms. Isabel joined the newly established Foreign Affairs Cadre Training School to engage in English teaching. In 1949, when the People's Republic of China was established, the Foreign Affairs Cadre Training School was moved to Beijing, and it was later renamed the Foreign Language School (now Beijing Foreign Studies University). From that point, Ms. Isabel has remained in China, and is a pioneer in English teaching in New China. On September 17, 2019, Chinese President Xi Jinping signed the President's Order to grant Isabel, a Canadian friend and veteran expert of Beijing Foreign Studies University of China—the Medal of Friendship of the People's Republic of China.

尽管2号小洋楼没有被拆除，但为了让更多人居住，阴台都封了做房间使用。这是工人们正在修缮2号楼屋顶的瓦片，俗称"捡瓦"。戚亚男摄于2009年。

Although the No. 2 small house has not been demolished, in order to let more people to live, the long corridor has been sealed for use as rooms. Workers are repairing the tiles on the roof of Building 2, commonly known as "picking up tiles". Photo by Qi Yanan in 2009.

饶和美一家离开2号小洋楼后，苏继贤一家又搬了进来。苏继贤是加拿大人，他毕业于维多利亚大学，1908年来四川传教。他是一位建筑工程师，主要从事教会建筑的修建。1925年苏继贤来到华西协合大学负责大学校园的校舍修建与

239

维修并担任文科教员，直到1950年7月才离开成都回国。华西坝上的化学楼、大学医院、连接医牙学院中间的部分建筑等都是苏继贤负责监督施工的。说一口流利四川话的苏继贤平易近人，当年他教过的学生郭祝崧回忆："他（苏继贤）经常对别人说，'叫我工程师就见外了，够朋友，喊我苏木匠，亲热些！'"苏继贤的子女都是在华西坝出生的，长大之后回加拿大完成大学学业。他的大儿子苏维廉1917年在四川乐山出生，他从加拿大多伦多大学毕业后，于1941年回到华西协合大学担任财务主管直到1952年离开成都回加拿大，回国后他一直致力于加中友好关系的促进工作，他是加拿大加中友好协会的创始人之一。二儿子苏约翰1919年出生在成都，16岁回加拿大继续学习。1941年，22岁的苏约翰参加加拿大皇家海军，直到1945年二战胜利后到多伦多大学读书。毕业后苏约翰从事外交工作，1972年到1976年他出任加拿大第二任驻华大使。

After the Brown family leaving the No. 2 small house, the Walter. G. Small family moved in. Mr. Walter. G. Small, a Canadian, graduated from the University of Victoria and came to Sichuan to preach in 1908. He was an construction engineer mainly engaged in the construction of church architecture. In 1925, Mr. Small came to West China Union University to take charge of the construction and maintenance of the university buildings on university campus and served as a liberal arts teacher, until leaving Chengdu in July 1950 to return home. Mr. Small was responsible for the construction of the buildings in Huaxiba, such as the Chemical Building, the University Hospital and the middle section connecting the Medical and Dental College were constructed under his responsible. Mr. Small speaks fluent Sichuan dialect, and is very approachable. Mr. Guo Zhusong, a student he taught, recalled, "He often said to others, 'If you call me an engineer, you are a stranger. If you are a real friend, call me carpenter Su, that's more intimate!'" Mr. W.G. Small's children were brought up in Huaxiba, Sichuan Province, and then returned to Canada to finish their college studies. His eldest son, W.W. Small, was born in Leshan, Sichuan Province in 1917. After graduating from the University of Toronto, he returned to West China Union University in 1941 and served as treasurer until 1952, when he left Chengdu for Canada. After returning to Canada, he devoted himself to promoting the friendly relations between Canada and China. He was one of the founders of the Canada-China Friendship Association. Mr. W. G. Small's second son, J. Small, was born in Chengdu in 1919 and returned to Canada to continue his study at the age of 16. In 1941, at age 22, J. Small joined the Royal Canadian Navy and went to the University of Toronto to study after the victory in World War II in 1945. After graduation, he pursued diplomatic work, serving as Canada's second ambassador to China from 1972 to 1976.

1951年10月，生物学家刘承钊教授出任校长直至1968年，入住2号小洋楼。关于刘承钊的内容参见前文。

1976年刘承钊去世。在这之后，2号小洋楼就分配给学校的教职工居住一直到现在。

In October, 1951, the biologist Professor Liu Chengzhao moved into the No. 2 Small House.

He served as the president of the university until 1968 (See above).

Professor Liu Chengzhao died in 1976. After this, the No. 2 small building has been assigned to the university staff residence until now.

抗战时期成都对外文化交流的国际窗口——东西文化学社
East-West Culture Society, the International Window of Cultural Exchange During the Counter-Japanese War in Chengdu

华西坝上修建的众多小洋楼几乎都是供坝上的外籍教师居住的，但在20世纪40年代华西坝上有栋小洋楼却成为一个社会团体——"东西文化学社"的办公场所，从事国际文化交流活动，该学社是由华西协合大学文学院罗忠恕院长创办。

There were many small foreign style houses in Huaxiba. Almost all of the small houses were for the foreign teachers of the university to live. But in the 1940s, there was a small house in Huaxiba that became the office of a social group—The East and West Cultural Society, engaged in international cultural exchange activities. The society was founded by Dr. Luo Zhongshu, Dean of the School of Literature, West China Union University.

1937年，罗忠恕赴英国牛津大学留学。刚到伦敦他就参加了在牛津大学召开的世界基督教大会。在会上，他把我国孔子主张的大同世界和希腊柏拉图的理想做了比较，引起了大家的关注。在牛津大学学习期间，罗忠恕在英国很多大学做中国文化的演讲，和那里的师生交流、讨论国际问题，特别是中国的抗日战争问题。因此时常有不少学者请他去作学术演讲和交流，由此还引起牛津大学和剑桥大学成立了一个推动中英两国文化交流的"中英文化合作委员会"。

In 1937, Dr. Luo Zhongshu went to Oxford University in England to study. When he just arrived in London, he attended the World Christian Conference held at the University of Oxford, where he compared the ideal of Datong (Great Harmony) World advocated by Confucius in China with the ideal of Plato in Greece, which attracted everyone's attention. During study in Oxford, Dr. Luo Zhongshu lectured on Chinese culture in many universities in the UK, communicated with teachers and students there, and discussed international issues, especially China's Counter-Japanese War. Therefore, as a result, many scholars often invited him to give academic lectures and exchanges, which caused the University of Oxford and Cambridge to set up a "China-British Cultural Cooperation Committee" to promote cultural exchanges between China and the UK.

1940年，罗忠恕回国。当时正值抗战时期，国内高校内迁西南，金陵大学、金陵女子文理学院、齐鲁大学、燕京大学等先后迁到成都华西坝。罗忠恕联络各大学专家、学者以及社会名流，于1942年11月19日成立了一个研讨国际文化交流的"东西文化学社"，为国内学术界进行国际交流开了一扇窗口。历史学家钱穆为此写了《东西文化学社缘起》文章，文中写道："罗君忠恕，游学海外，有心此事，曾于民国二十八年之冬季，两次在英伦牛津、剑桥两大学发表其对东西两大民族应对双方文化合作更进一步之发挥与相互融贯之工作之演讲，颇蒙彼中有识者之同情，并在牛津、剑桥两大学成立中英学术合作委员会，且发表宣言，赞同此事。此外国际知名学者如爱因斯坦、杜黑舒、怀特黑、杜威、罗素诸氏，均通函问，愿赞斯举。罗君返国，因发表中国与国外大学学术合作之建议一小册，略道其梗概，同人等对罗君意见甚表赞同，因感其共组学会，共同努力之必要，遂发起东西文化学社，草拟简章。采此广征国内同志集力进行，一面拟约请国外学者密切联系，共同合作。"

In 1940, Dr. Luo Zhongshu returned home. At that time, during the Counter-Japanese War, domestic universities moved to the southwest, University of Nanking, Ginling College, Cheeloo University, Yenching University, etc., successively moved to Huaxiba, Chengdu. Dr. Luo Zhongshu contacted with experts, scholars and celebrities from various universities, and on November 19, 1942, set up an "East-West Cultural Society" to discuss international cultural exchanges, which had opened a window for domestic academic circles to conduct international exchanges. Historian Qian Mu wrote in the article *The Origin of the East—West Cultural Society*, "When Luo Zhongshu studied in the UK, he planned to establish an academic institution to promote cultural exchanges between China and the UK. In 1939, Luo Zhongshu delivered two speeches at Oxford and Cambridge universities. He said that the two peoples of the East and the West should have cultural exchanges. Luo Zhongshu's speech was recognized by many scholars. Then they established a China-British Cultural Cooperation Committee at Oxford and Cambridge universities and issued a declaration. In addition, well-known international scholars such as Einstein, Driesch, Whitehead, Dewey and Russell all wrote to praise the matter. After Luo Zhongshu returned home, he established the East West Culture Society with professors from Huaxiba five universities. It has opened a window for domestic academic circles to conduct international exchanges. After Mr. Luo returned to China and published a small book of suggestion on academic cooperation between China and foreign universities.His colleagues agreed with his opinions and felt that it was necessary to form a society and work together. So Mr. Luo set up the East West Cultural Society and drafted a brief chapter to recruit domestic comrades,while planning to invite foreign scholars contact closely and cooperate with each other."

学社成员大都是华西坝上五所大学各学科的专家和教授，有钱穆、蒙文通、施友忠、何鲁之、姜蕴刚、倪青原、罗忠恕、何文俊、冯友兰、萧公权、萧一山、冯汉骥、李安宅、汤藤汉、刘国钧、闻宥、李珩、董时进、常乃德、侯宝璋、罗念生、郑集、陈钟凡、王绳祖、吴其玉、吴俊升、郑德坤、唐君毅、贺昌群、牟宗三、郭本道、吕叔湘、蒙思明等40多位学者，以及各界社会名流。

学社社长为罗忠恕，聘请孔祥熙、张群、张嘉璈、孙哲生和顾维钧为名誉社长。

Most of the members of the society were experts and professors in various disciplines of the five universities in Huaxiba. More than 40 scholars, and noted public figures were the members. They were Qian Mu, Meng Wentong, Shi Youzhong, He Luzhi, Jiang Yungang, Ni Qingyuan, Luo Zhongshu, He Wenjun, Feng Youlan, Xiao Gongquan, Xiao Yishan, Feng Hanji,

Li Anzhai, Tang Tenghan, Liu Guojun, Wen Yu, Li Heng, Dong Shijin, Chang Naide, Hou Baozhang, Luo Niansheng, Zheng Ji, Chen Zhongfan, Wang Shengzu, Wu Qiyu, Wu Junsheng, Zheng Dekun, Tang Junyi, He Changqun, Mou Zongsan, Guo Bendao, Lü Shuxiang, Meng Siming and others.

The society president was Luo Zhongshu, and Kong Xiangxi, Zhang Qun, Zhang Jiaao, Sun Zhesheng and V.K. Wellington Koo were honorary presidents.

罗忠恕还聘请了五所大学校长张凌高、梅贻宝、汤吉禾、吴贻芳和陈裕光，四川大学校长黄季陆，以及四川省教育厅厅长郭有守为名誉社员。

Dr. Luo Zhongshu also invited the presidents of the five universities, namely, Zhang Linggao, Mei Yibao, Tang Jihe, Wu Yifang and Chen Yuguang, Huang Jilu, President of Sichuan University, and Guo Yousheng, Director of Provincial Department of Education, as honorary members.

成都军界24军军长刘文辉专门提供了一笔费用，在华西后坝101号（现人民南路三段19号）购买了一栋小洋楼作为学社的社址。学社定期在这里召开学术座谈会，并在此接待来参加学社讲学的中外学者，国内的如张东荪、梁漱溟等，国外的有李约瑟、艾格斯顿等。

Liu Wenhui, Commander of the 24th Army of Chengdu, provided a fee to buy a small house at No. 101 of Huaxihouba (now No. 19 of South Renmin Road) as the site of the society. The society regularly held academic symposiums here and received Chinese and foreign scholars to attend the lectures of the society, such as Zhang Dongsun and Liang Shuming in China, and Joseph Needham and Aigston abroad.

1942年12月3日，中央社的一则新闻"英议会访问团艾尔文爵士，劳森先生及顾大使等，今日下午四时半赴华西坝参观……同时卫德波先生则在华西坝对四千余学生讲演'战后的世界'均受到热烈欢迎，五时出席中西文化社茶会……"揭开了学社进行中外文化交流的开始。就在学社刚成立十来天，驻英大使顾维钧就陪同英国议会访华团从重庆来成都访问。他们参观了华西坝五大学，作讲演，与学社成员交流，观看中英两国足球比赛。

On December 3, 1942, a piece of news from China Central News Agency told, "The British Parliament delegation Sir Lord Aiwyn, Mr. J. J. Lawson and Ambassador Gu visited Huaxba at 4:30 p. m. today...At the same time Mr. H. J. Scry Ageour Wedderburn, gave a speech on 'The World after the War' to more than 4,000 students in Huaxiba, which was warmly welcomed. At 5 o'clock, they attended the tea party of the Chinese and Western Cultural Society..." opened the beginning of cultural exchanges between China and foreign countries of the Society. Just ten days after the establishment of the Society, Ambassador to the UK V.K. Wellington Koo accompanied the British Parliamentary Delegation to China from Chongqing to visit Chengdu. They visited Huaxiba Associated Universities in Chengtu, gave lectures, communicated with members of the society and watched football matches between China and Britain.

访华团成员卫德波在讲演中赞扬中国人民在抗日战争中的英勇抗战，为全体英国人民钦佩。他力主中英战后长期合作，不仅政治上成立同盟，商业上尤需发展，中英永远为世界和平而努力。

此后，在罗忠恕社长的带领下，东西文化学社进行了一系列国际学术交流活动，为成都打开了一扇和国际对话的窗口。

In his speech, Wedderburn, a member of the visiting delegation to China, praised the Chinese people's heroic war of resistance Counter-Japan over the past five years and was admired by all the British people.

Since then, under the leadership of President Luo Zhongshu, the East and West Cultural Society had held a series of international academic exchange activities, opening an international dialogue window for Chengdu.

尽管东西文化学社使用过的小洋楼很早就被拆除了，但2014年东西文化学社旧址被选为成都首批50个文化地标之一，并挂牌向市民展示这里曾经是抗战时期成都对外进行文化交流的国际窗口。

Although the small houses of the East and West Cultural Society was demolished long ago, in 2014, the former site of the East and West Cultural Society was selected as one of the first 50 cultural landmarks in Chengdu and listed to show the public that it was once an international window of cultural exchanges between Chengdu and the outside world during the Counter-Japanese War.

"第十一号住宅的耗子"
"The Mice of No. 11"

"轻轻的敲门声刚刚响过，书房门立即打开了。为数八九人的学生鱼贯而入，他们都是二十几岁的男女青年：农科学生王宇光、燕京大学新闻系的李肇基、金陵大学的谢韬、华西大学的贾唯英、金陵女大的杨廷英，他们是共产党领导下的新民主主义青年团在成都的政治领导人。文幼章只知道他们属于'金陵大学星星团契'；像往常一样，盛了一大碗花生并用开水沏茶招待他们。当这些学生传递着花生吃时，他们开玩笑地说，他们是'十一号住宅的耗子'。"

"The study door swung open after only the faintest knock. A group of eight or nine students filed in, men and women in their early twenties: Wang Yu-guang, a student in agriculture, Li Chao-chi, Yenching school of journalism, Hsieh Tao from Nanking University, Chia Wei-yin of West China Union, Yang Tin-yin from Ginling Women's College. They were the political leaders in Chengtu of the New Democratic Youth League, an underground student organization led by the Communist Party. Endicott, who knew them only as 'the Sparks(Xing Xing Tuan) of Nanking University,' made them welcome as usual with hot water for tea and a large bowl of peanuts. As the nuts were passed around, the students joked that they were 'the mice of No. 11.'"

"星星团契成员最初接近文幼章是在听了他的几次公开布道和演讲之后，他们借口需要一个安全的地方来读某些著作——西方马克思主义的经典著作和来自延安的文章——如果在学生宿舍发现了这些东西,学生将被立即被捕。由于文幼章当过蒋介石的顾问,而且大家又知道他在教省长张群学英语,他们便认为他的家可以成为一个庇护所。'我们一些学习小组可以在你家开会吗?'他们问道。"

"The Sparks had originally approached Endicott after hearing some of his outspoken sermons and speeches, saying that they needed a safe place to study certain banned texts—the classics of Western Marxism and articles that came from Yenan—which, if found in the student dormitories, made the owner subject to immediate arrest. Since he had been an adviser to Chiang Kai-shek and because he was known to be tutoring Governor Chang Chun in English, they thought that his home might offer a sanctuary. 'Could some of our study groups meet at your house?' they had asked him."

"文幼章同意了,给他们一把房门钥匙,有时他参加讨论。在把窗帘紧闭之后,学生们揭开一片活动地板,取出并分发供讨论的阅读材料。文幼章听着他们谈话。"

"Endicott had agreed, giving them a key to his door, and sometimes he joined in their discussions. After making sure the curtains were drawn, the students would lift up a loose board in the floor and distribute the controversial reading material. As Endicott listened to them talk."

这是文幼章的传记《第十一号住宅的耗子》一章的开头,描述了一群华西坝五大学的进步学生来到他们的导师文幼章教授的小洋楼进行地下革命活动的情景。文幼章是华西协合大学的教授,是"中国人民友好使者",也是加拿大传教士。这部传记名叫《文幼章传——出自中国的叛逆者》,由其儿子文忠志所著。

This is the begins of the chapter *The Mice of No. 11*, the biography of James G. Endicott, which described a scene a group of progressive students from Huaxiba Associated Universities in Chengtu went to the small foreign style houses, residence of their advisor, Professor James G. Endicotd to conduct underground revolutionary activities. James G. Endicotd was a professor of West China Union University and was the "People's Friendly Messenger" in China, also a Canadian missionary. The biography was named *James Gareth Endicott—Rebel out of China* written by his son Stephen Endicott.

文幼章1898年生于四川乐山,其父文焕章1895年从加拿大来到四川传教。27年后,文幼章受父亲影响于1925年也来到四川重庆传教。1940年文幼章来到成都在华西坝上从事英语教学,文幼章的直接英语教学模式深受学生欢迎,其强调培养学生的英语阅读和听说能力。1932年,中华书局出版发行了文幼章撰写的《直接法英语读本》教科书。

James G. Endicott was born in Leshan, Sichuan Province in 1898. His father, J. Endicott came to Sichuan as a missionary from Canada in 1895. 27 years later, James G. Endicott was influenced by his father and also preached in Chongqing, Sichuan Province in 1925. In 1940, James G. Endicott came to Chengdu to engage in English teaching in Huaxiba. The direct english teaching mode of James

G. Endicott was very popular with students. The mode emphasizes on cultivating students' English reading and listening and speaking ability. As early as 1933, Zhonghua Book Company published the textbook of *Direct Method English Reader* written by James G. Endicott.

文幼章在四川这10多年里除了教书外，还投身到中国人民的抗日战争中去。1939年文幼章出任蒋介石的政治顾问，并为宋美龄发起的"新生活运动"担当社会顾问。

In addition to teaching in Sichuan for more than ten years, Mr. James G. Endicott also devoted himself to the Chinese people's Counter-Japanese War. In 1939, Mr. James G. Endicott served as a political adviser to Chiang Kai-shek, and served as a social adviser to the "New Life Movement" initiated by Ms. Soong Meiling.

文幼章在华西坝教书期间经常被各大学的学生团体邀请去发表演讲，他批评国民党的一党独裁和专政，支持学生提出的反对内战主张。

During his teaching in Huaxiba, James G. Endicott was often invited by student groups in various universities to give speeches. He criticized the one party dictatorship of the Kuomintang and suppressed the people, and supported the students' claims to against the civil war.

1945年抗战胜利后，云南昆明几个大学的师生在西南联合大学召开反内战的时事讨论会，国民党制造了震惊中外的"一二·一"惨案。华西坝上的五大学和国立四川大学的学生准备12月9日（星期天）在华西坝校园里先举行追悼在"一二·一"惨案中遇难的教师和同学、坚决反对内战、争取民主的大会，然后游行到市中心的少城公园（现在的人民公园）并发表演讲声援昆明学生的活动，学生代表邀请文幼章到场演说。文幼章虽然对国民党政府不满，但他之前却从未公开表示过反对。这一次他不顾自己的生命安危，答应了学生的请求，公开作演讲，表示了对学生运动的同情。

After the victory of the Counter-Japanese war in 1945, teachers and students of several universities in Kunming, Yunnan Province held a current affairs seminar against the civil war at the National Southwest Associated University. The Kuomintang created the "12·1" Tragedy that shocked China and the world. Students from the Associated Universities in Chengtu and the National Sichuan University were going to hold a meeting on December 9 (Sunday) in Huaxiba campus to mourn for the dead students in Kunming, resolutely opposed the civil war and strove for democracy, and then march to Shaocheng Park in the city center (now People's Park) and delivered a speech to support the activities of Kunming students. The student representative invited Mr. James G. Endicot to make a speech. Although Mr. James G. Endicott was dissatisfied with the Kuomintang government, he never publicly expressed his opposition. This time, he defied the wishes of the authorities and ignored his life and security. He promised the students' request gave a public speech and expressed his sympathy for the student movement.

那天在少城公园聚集了5 000多名市民，文幼章看见讲台四周站了30个头戴钢盔、身穿军服的警察。当文幼章一走上讲台后，台下立即爆发出一阵雷鸣般的欢呼声，这时一个穿军服的人把一颗没有拉引线的手榴弹扔上讲台以此恐吓。文幼章毫不惧怕，他以流利的四川话向大家讲道："罗斯福总统提出租借法案以武器供给中国，是要帮助中国打倒日本法西斯，是要帮助建设一个四大自由的中国。拿武器来支持中国内战绝不是他的心愿。"台下的学生一致高呼"拥护罗斯福的进步政策，反对美国干涉中国的内政！" 1946年6月，文幼章因支持学生反对内战而被迫辞职回国。文幼章回到加拿大后致力于国际和平事业，1948年当选多伦多保卫世界和平委员会主席，1949年当选加拿大全国和平大会主席。

More than 5,000 citizens gathered in Shaocheng Park that day. Mr. James G. Endicott saw 30 policemen wearing helmets and military uniforms standing around the platform. As soon as Mr. James G. Endicott got on the platform, a thunderous cheer burst under the platform. At this time, a man in military uniform threw a grenade without a pull cord on the podium to intimidate him. Mr. James G. Endicott was not afraid. He spoke to everyone in fluent Sichuan dialect, "President Roosevelt proposed the lease and loan act to supply weapons to China to help China defeat Japanese fascism and help build a four freedom China. Taking arms to support the Chinese civil war was by no means his wish." The students in the audience unanimously shouted "support Roosevelt's progressive policy and oppose American interference in China's internal affairs!" In June, 1946, Mr. James G. Endicott was forced to resign and return home because he supported students against the civil war. After returning to Canada, he devoted himself to the cause of the international peace movement. He was elected Chairman of the Toronto Committee for the Defence of World Peace in 1948 and Chairman of the Canadian National Peace Conference in 1949.

1953年，文幼章荣获国际和平斯大林奖。

In 1953, Mr. James G. Endicott won the Stalin Peace Prize.

1956年，文幼章被《人民日报》称为"中国人民的老朋友"。

In 1956, Mr. James G. Endicott was called "an old friend of the Chinese people" by the *People's Daily*.

1965年，中国人民对外友好协会授予文幼章"人民友好使者"的称号。

In 1965, Mr. James G. Endicott was awarded the title of "People's Friendship Envoy" by the Chinese People's Association for Friendship with Foreign Countries.

2009年，在评选"100位为新中国成立做出突出贡献的英雄模范人物和100位新中国成立以来感动中国人物"中，文幼章被评为"致力于世界和平友好事业，世界著名的和平战士"。

In 2009, in the selection of "100 heroes and models who have made outstanding contributions to the founding of new China, and 100 people who have moved China since the founding of new China", Mr. James G. Endicott was rated as "dedicated to the cause of world peace and friendship, a world-famous warrior of peace".

1993年11月27日，文幼章去世，享年95岁。

On November 27, 1993, Mr. James G. Endicott died at the age of 95.

红色小洋楼——成都《挺进报》的故事
Red Small Foreign Style House, the Story of Chengdu XNCR

在华西坝上众多的小洋楼里有一栋很特别，它就是外籍教师云从龙的住宅。中华人民共和国成立时期，成都地下党办的一份报纸上面报道的很多消息都是从这栋小洋楼里的收音机里所获得，小洋楼的主人云从龙是华西坝上的3位红色传教士之一，因而该栋小洋楼就被称为红色小洋楼，即现在的人民南路三段16号。

Among the small foreign houses in Huaxiba, one was very special. It was foreign teacher Mr. L. E. Willmott' residence. The founding period of the People's Republic of China, a lot of news reported in a Chengdu underground party newspaper was obtained from the radio in this small house. Mr. L. E. Willmott, the host of the small house, was one of the three red missionaries in Huaxiba. Therefore, the small house was called a red small house. Now it was located at No. 16, section 3, South Renmin Road.

云从龙1895年出生在加拿大多伦多，大学毕业后于1921年到中国西部传教，在四川省仁寿县创办了华英中学（现仁寿一中）并出任校长。1932年，云从龙来到成都华西协合高级中学校担任校务长，之后他出任华西协合大学教务长，同时他还兼任华西协合大学财务主管，并担任"英语""教育"等课程的讲授。

L. E. Willmott was born in Toronto, Canada, in 1895. After graduating from university, he went to western China to preach in 1921. He founded Huaying Middle School (now Renshou No.1 Middle School) in Renshou County, Sichuan Province and served as the President. In 1932, Mr. L. E. Willmott came to Chengdu West China Union Senior Middle School as Provost. Later, he served as Provost of West China Union University. At the same time, he also served as Treasurer of the university and taught English, Education and other courses.

1946年文幼章辞职离开中国回加拿大后，云从龙接替文幼章担任金陵大学星星团契的顾问，他居住的小洋楼就成了团员的活动场所。当年在国统区阅读进步书刊是违法的，云从龙支持学生的课外阅读自由，他不仅把小洋楼的阁楼腾出来建了一间图书阅览室，还为学生购买了大量的进步书刊放在图书阅读室供学生阅读，有《论联合政府》《论共产党员修养》《中国土地法大纲》《小二黑结婚》《王贵与李香香》《解放日报》《大公报》等，使小洋楼成了名副其实的"红色图书馆"。他与华西坝上美国的费尔朴、英国的徐维理反对国民党的独裁和内战，支持学生运动，被称为传教士中的"三位布尔什维克"。

拾陆・华西坝上的教职员住宅——小洋楼

位于人民南路三段16号的小洋楼，云从龙的旧居，曾经的"红色小洋楼"。咸亚男摄于2003年。

The small house located at No. 16, Section 3, South Renmin Road, the former residence of L.E. Willmott, once the "red small foreign house". Photo by Qi Yanan in 2003.

After Mr. James G. Endicott resigned and left China to return to Canada in 1946, Mr. L. E. Willmott succeeded Mr. James G. Endicott as the adviser of "the Sparks (Xing Xing Tuan)" of university students. The small house where he lived became the activity place of the fellowship. In those years, it was illegal to read progressive books and periodicals in the Kuomintang controlled areas. Mr. L. E. Willmott supported students' free reading after class. He not only vacated the attic of the small house and built a book reading room, but also bought a large number of progressive books and periodicals for students to read in the reading room, including *On the Coalition Government, On the Cultivation of Communist Party Members, The Outline Land Law, Xiaoerhei's Marriage, Wang Gui and Li Xiangxiang, Liberation Daily, The Dagong Daily* and so on. The reading room had become a veritable "Red Library". He, along with D. L. Phelps of the U.S. and W. G. Sewell of the U.K., opposed the dictatorship and civil war of the Kuomintang and supported the student movement, thus being called the "Three Bolsheviks" among missionaries.

1947年，成都中共地下党决定由马识途来领导和筹办一份地下党报，报纸的宗旨很明确：就是把从收音机里收听到的解放军节节胜利的消息油印成报纸，再散发到党组织成员和进步群众中去。报纸的名称就用延安广播电台的呼号"XNCR"。

In 1947, the Chengdu Underground Communist Party decided

that Mr. Ma Shitu would lead and organize an Underground Party newspaper. The purpose of the newspaper was very clear, to print the news of the PLA's continuous victory received from the radio into a newspaper, and then distribute it to the party organization and the progressive masses. The name of the newspaper used the Yan'an Radio Station's call sign "XNCR".

马识途找到了中共四川大学支部书记王放，她的公开身份是国立四川大学历史系的学生，由她来具体负责办这份报纸。这是一份现在鲜为人知的成都地下党报，它未遭到国民党特务的破坏，那是因为办报人的胆大心细，绕过了国民党特务设置的种种陷阱、关卡和控制，并得到了国际友人云从龙的无私帮助，使《XNCR》能源源不断地将解放军的捷报及时传到党员和进步群众中去。这不仅鼓舞和激发了大家的斗志，也提高了群众对当时局势的认识。

Mr. Ma Shitu approached Miss Wang Fang, the branch secretary of the CPC Sichuan University, publicly identity known as a student in the History Department of the National Sichuan University. She was responsible for running the newspaper. Now this is a little-known Chengdu Underground Party newspaper. At that time, it was not destroyed by the Kuomintang spy, because of their boldness, passing around the traps, checkpoints and controls set up by the Kuomintang spy and got the international friends Mr. L. E. Willmott selfless help. So that *XNCR* could continuously spread the PLA victory of the news timely to the Party members and the progressive masses. This not only inspired everyone's morale, but also raised the people's awareness of the current situation.

在国统区要想办一份红色报纸是很不容易的，国民党特务机关把办报的资源全部掌控了，不要说买不到短波收音机，就连买一部普通的收音机都要被国民党特务跟踪调查。没有短波收音机，就收不到延安的消息；没有消息，就无法办报纸。但这难不倒马识途和王放，他们自己动手组装了一部短波收音机。有了收音机，就有了消息的来源，这样一张红色报纸《XNCR》就成功诞生了，具体负责工作的就只有王放一人，收听并记录延安广播、刻写蜡纸、油印、分发等一系列工作几乎全由她一人包干了。《XNCR》不断登出解放军打胜仗的消息以及中共中央的一些文件、评论、声明等文章。该报3天一期，若有大胜利，还要加印号外。

It was not easy to run a red newspaper in the Kuomintang ruling area. The Kuomintang secret agencies controlled all the resources of the newspaper. Not only people could not buy a short-wave radio, even buying an ordinary radio was hard. Because Kuomintang secret spies always tracked and investigated any radio buyer. Without a short-wave radio, they could not receive the news from Yan'an, and no news, they could not run a newspaper. But this was not difficult for Mr. Ma Shitu and Miss Wang Fang. They assembled a short-wave radio. With the radio, there was a source of the news. Such a red newspaper *XNCR* was successfully born. Only Miss Wang Fang was responsible for the specific work. She listened to Yan'an radio and made notes. Almost all the series of work such as engraving wax paper, oil printing and distribution and so on was responsible by her. *XNCR* kept making news of the PLA's victory, as well as documents, comments, statements from the

Central Committee of the Communist Party of China. The newspaper was published every three days. If there was a big victory, it would printed an extra edition.

时间一久，国民党特务终于注意到了《XNCR》，马识途他们办报的地方被特务注意到了，他们只能转移走，但在什么地方收听广播成了继续办报的难题。

After a long time, the spy finally noticed *XNCR*. The place where Mr. Ma Shitu ran the newspaper was noticed by the spy, and they had to move away, but where to listen to the radio to continue running the newspaper become the key problem.

当时马识途就是通过云从龙的帮助来到华西协合高级中学校教英文，以教书作为职业掩护他从事的地下工作，云从龙是知道马识途在从事地下工作的。因为在收听延安广播上遇到了困难，《XNCR》报纸就有办不下去的可能。马识途决定带王放去求助云从龙。当知道他们的来意，并了解到马识途与王放的特殊关系后，云从龙夫妇热情地接待了他们，且欣然接受了他们的请求。云从龙有一台落地大收音机，灵敏度非常高，选择性也好，王放一下就能收到延安广播的消息。当即约好除周六和周日外，他们每天晚上到云从龙家客厅里收听两个小时的延安广播，对外名义是学习英语。几天后，马识途因还要忙于其他工作，此后一直就是王放一个人在负责了。云从龙夫妇对王放十分关照，常常端一小碟点心让她吃。

At that time, Mr. Ma Shitu came to West China Union Senior High School to teach English with the help of Mr. L. E. Willmott. He took teaching as a profession to cover up underground work. Mr. L. E. Willmott knew that Mr. Ma Shitu was engaged in underground work. Now, because Mr. Ma Shitu and his comrades had encountered difficulties in listening to Yan'an radio, *XNCR* newspaper might not be able to run. Mr. Ma Shitu decided to take Miss Wang Fang to ask Mr. L. E. Willmott for help. When Mr. L. E. Willmott knew their intention and understood the special relationship between Mr. Ma Shitu and Miss Wang Fang, Mr. L. E. Willmott couple warmly received them and gladly accepted their request. Mr. L. E. Willmott had a large landing radio with very high sensitivity and selectivity. Miss Wang Fang could catch the news of Yan'an once she played the radio. They immediately made an appointment. Except Saturday and Sunday, they went to his living room every night to listen to two-hour Yan'an news from the radio, in the name of learning English. A few days later, Mr. Ma Shitu was busy with other work. Since then, Miss Wang Fang had been working alone. Mr. L. E. Willmott couple were very considerate to Miss Wang Fang and often bring her a small plate of snacks.

云从龙帮助中共地下党办《XNCR》报非常不易，这使马识途终生难以忘怀这份不可多得的友谊。2008年4月，云从龙在成都仁寿出生的儿子云达乐从加拿大回到成都寻旧，马识途专门以家宴的形式宴请云达乐。席间马识途向在座的宾客讲述了当年云从龙夫妇是如何帮助地下党办报的故事。马识途还把他写的描述当年他从事地下党工作的纪实小说《在地下》送给云达乐，并在书的扉页上写道："您的父亲云从龙先生给中国人民解放斗争的热情帮助以及我们在华西协中所建立的深厚友谊，是永远不会忘记的记忆。"

Mr. L. E. Willmott helped the Underground Party of the Communist Party of China to run *XNCR* newspaper, which made Mr. Ma Shitu unforgettable for life. Mr. Donald E. Willmott is the son of Mr. L. E. Willmott. Mr. Donald E. Willmot was born in Renshou, Chengdu. In April, 2008, he returned to Chengdu from Canada to look for the old time. Mr. Ma Shitu specially entertained Mr. Donald E. Willmott in the form of a family banquet. During the dinner, Mr. Ma Shitu told the guests how the L. E. Willmott couple helped the underground party run newspapers. Mr. Ma Shitu also gave Mr. Donald E. Willmott his documentary novel *Underground* describing his underground work in those years, and wrote on the title page of the book, "Your father, Mr. L. E. Willmott, warm help to the Chinese people's liberation struggle and the deep friendship we established in the West China Union Middle School will never be forget."

2015年，这栋小洋楼被确定为成都市文化地标并挂牌保护。

In 2015, the small house was identified as a cultural landmark of Chengdu city and listed for protection.

华西坝老建筑的前世今生
The Stories of the Historic Architecture of Huaxiba

拾柒

中加两国人民友谊的见证

志德堂

The Canadian School in Huaxiba, Witnessing the Friendship Between Chinese and Canadian People

加拿大学校（简称CS学校）大楼1915年动工，1924年竣工，该楼是来四川的传教士的子女寄宿、读书的地方，CS学校是一所从幼儿园到高中的全日制学校。1909年3月9日，加拿大传教会在四圣祠北街教堂后面的一间房子里开办了加拿大学校，当时只有5位学生，CS学校大楼在华西坝落成后该校就从四圣祠北街迁来了。该楼一层为防潮地下室，上面两层才是供教学办公使用的，两层正面都有长长的阴台，墙体是西式拱券的结构，刚竣工时屋顶并非用青瓦而是用白铁皮覆盖，上面有壁炉烟囱和透气的老虎窗，在以后的维修中不仅楼层增加了一层，变成了三层楼房，阴台也装上窗户当房间使用，而且屋顶上的白铁皮也被大红瓦所取代。

The construction of Canadian School (CS for short) building was started in 1915 and completed in 1924. The building was a boarding school for the children of missionaries coming to Sichuan. It was a full-time school from kindergarten to high school. On March 9, 1909, the Canadian missionary church opened a Canadian School in a house behind the church on the Sishengci North Street. At that time, there were only five students. After the CS building was completed in Huaxiba, the school moved from Sishengci North Street to the building in Huaxiba. The first floor of the building was a moisture-proof basement, and the upper two floors were used for teaching and office. There were long shade platforms on the front of both floors and the walls were western-style arch structure. When it was just completed, the roof was not covered with grey tiles, but with white iron sheet, with fireplace chimney and roof ventilation windows on it. In late maintenance, not only one floor was added to turn into a three-story

加拿大学校大楼落成，开始接纳孩子们入住，摄于1920年4月23日。图片来源：http://loonfoot.com/Leonard/LeonardScrapbook/China/School/04.htm。

The Canadian School building was completed and the children were admitted. Photo taken on April 23, 1920. Photo source: http://loonfoot.com/Leonard/LeonardScrapbook/China/School/04.htm.

原加拿大学校大楼后来称为第七教学楼，现在被命名为志德堂，是华西公共卫生学院的行政办公楼。戚亚男摄于2021年。

The former CS building, later known as the Seventh Teaching Building, and now also named Zhide Hall, is the administrative office building of the West China School of Public Health. Photo by Qi Yanan in 2021.

building, but also the long shade platforms were equipped with windows for use as room, and the white iron sheet on the roof was replaced by big red tiles.

抗战时期，CS学校的师生为了躲避日机的轰炸外迁成都郊县仁寿，CS学校大楼就借给金陵女子文理学院使用，成为该校的第二宿舍。

During the Counter-Japanese War, the teachers and students of CS moved to Renshou, a suburb county of Chengdu, to avoid the bombing of Japanese planes. The CS building was lent to Ginling College and became the second dormitory of the College.

1950年加拿大学校关闭，不久该大楼更名为第七教学楼，之后全国高等院校调整，华西大学由综合性大学改为以医药为主的四川医学院，学院建立了卫生系，第七教学楼就作为卫生系行政、教学、科研的场所，被誉为中国公共卫生之父的陈志潜也调到卫生系执教，直到2000年去世。

In 1950, the Canadian School was closed, and soon the building was renamed the Seventh Teaching Building. After the adjustment of the departments of colleges and universities nationwide, the university was changed from a comprehensive university to Sichuan Medical College focusing on medicine. The College established the Department of Health, and the Seventh Teaching Building was the place for administration, teaching and scientific research of the Department of Health. Dr. Chen

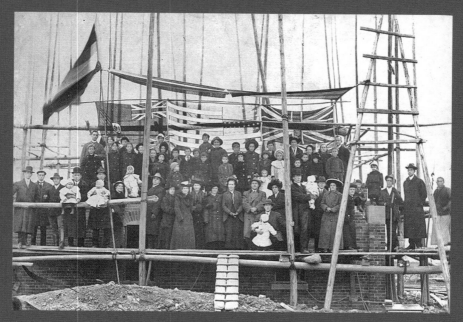

上图：1915年12月17日，加拿大学校大楼安放奠基石时，传教士们带着他们的孩子在大楼地基前合影留念。脚手架上悬挂着加拿大、美国、英国的国旗，右边第3位是该楼的建筑工程师能贤哲，他手上还握着刚把奠基石吊上去安置在外墙体上的绳子。

图片来源：http://loonfoot.com/Leonard/LeonardScrapbook/China/School/01.htm

On December 17, 1915, when the cornerstone was placed in the Canadian School building, missionaries took a photo with their children in front of the foundation of the building. The flags of Canada, the United States and Britain were hung on the scaffold. The third on the right was Mr. William M. Leonard, the construction engineer of the building. He was still holding the rope that had just hoisted the cornerstone up to the outer wall.

Photo source: http://loonfoot.com/Leonard/LeonardScrapbook/China/School/01.htm

下图：2016年11月9日，华西加拿大学校陈列馆在志德堂举行了开馆仪式，陈列馆落户志德堂是中加两国人民友谊的象征。戚亚男摄于2016年。

On November 9, 2016, the opening ceremony of West China Canadian School Museum was held in CS building. The establishment of the museum in CS building is a symbol of the friendship between Chinese and Canadian peoples. Photo taken by Qi Yanan in 2016.

Zhiqian (C. C. Chen), known as the father of China's public health, was also transferred to the Department of Health of the College to teach until his death in 2000.

加拿大学校大楼现为四川大学公共卫生学院的教学楼。2013年5月，该楼被命名为志德堂，"志"即陈志潜名字中的"志"。

The Canadian School building is now the teaching building of the School of Public Health of Sichuan University. In May 2013, the building was named Zhide Hall, which has the same "Zhi" with "Zhi" in Chen's name.

2014年9月，志德堂被列入成都市历史建筑保护名录。

In September, 2014, The Canadian School building was listed in the Historic Building Protection List of Chengdu.

2016年11月，在志德堂里利用楼梯、走廊的墙壁和两间教室建了华西加拿大学校陈列馆。

In November, 2016, the West China Canada School Exhibition Hall was built in the Canadian School building by using stairs, corridor walls and two classrooms.

传播中国文化的加拿大学校校长黄思礼
Mr. Lewis C. Walmsley, the Principal of Canadian School, Spreading Chinese Culture

黄思礼是加拿大学校的第二任校长，他1897年出生于加拿大安大略省。黄思礼在读多伦多大学维多利亚学院期间结识了一位女同学黄素芳，两人相爱了。黄素芳的父亲启尔德是中国西部第一所西医诊所和华西协合大学的创办人之一，她是在成都出生和长大后回加拿大接受高等教育的。大学毕业后，1921年黄思礼与黄素芳结婚后两人双双来到四川从事传教工作。两人在成都学习了两年中文后，黄思礼被任命为CS学校校长，黄素芳在该校担任教师。

Mr. Lewis C. Walmsley, the second principal of the Canadian School, was born in Ontario, Canada, in 1897. When he studied at Victoria College at the University of Toronto, he met Miss Constance Kilborn, a student of the University, and they fell in love. Miss Constance Kilborn's father, O. L. Kilborn was one of the founders of the first western medicine clinic in western China and West China Union University. Miss Constance Kilborn was born and raised in Chengdu then back to Canada for higher education. After graduating from college, Mr. Walmsley married Miss Constance Kilborn and both came to Sichuan to do missionary work in 1921. After the two studied Chinese for two years in Chengdu, Mr. Walmsley was appointed principal of the CS, where Ms. Constance Kilborn worked as a teacher.

加拿大学校是一所专门为在四川的传教士子女开办的学校，学校的教学方式也

是西方的，但黄思礼他们把中国的传统文化融入教学当中去，让这些外国孩子从小就接触和了解中国文化。为了让孩子们学到正统的中国文化，黄思礼专门聘请中国名家、名师来CS学校任教。1938年，擅长中国画及书法的国立中央大学教授、教育部美术教育委员会委员许士骐因南京沦陷流亡到四川，黄思礼立刻聘请他担任CS学校艺术及中国文化史导师，让他教孩子们画国画，给孩子们上中国文化的课程。

The Canadian School was a school for the children of missionaries in Sichuan. The teaching method was also in a western way. But Mr. Walmsley also integrate traditional Chinese culture into teaching, so that these foreign children could contact and understand Chinese culture from an early age. In order to let the children learn the orthodox Chinese culture, he specially hired Chinese masters and famous teachers to teach in CS. In 1938, Mr. Xu Shiqi, a professor of the National Central University who specialized in Chinese painting and calligraphy and was a member of the Fine Arts Education Committee of the Ministry of Education, went into exile in Sichuan for the fall of Nanjing. Mr. Walmsley immediately hired him as the tutor of art and Chinese culture history at CS, asking him to teach CS children to draw traditional Chinese painting and give them lessons in Chinese culture.

1933年成都加拿大学校的伊莎白（前排右）等4位高中毕业生与校长黄思礼（后排中）在志德堂侧翼入口处合影。图片来源：柯马凯提供。

In 1933, four high school graduates of CS, including Isabel (right, front row), took a photo with principal Walmsley (middle, back row) at the wing entrance of CS. Photo source Michael Crook provided.

在这种氛围影响下,CS学校的孩子们对中国文化有了自己的理解,在他们的笔下如此生动地描写道:

Under the influence of this atmosphere, CS children had their own understanding of Chinese culture, so vivid description in their pen:

> 在中国到处可以看见线条优美、体态匀称的龙。它被画在丝绸和瓷器上,被编织进锦缎里,被雕刻在木头上,被绣在绸缎上,被铸造在青铜器中,被凿刻在大理石上。中国民俗中充满了数不清的龙的神奇传说。皇上的王座叫作"龙椅",皇上用的毛笔叫"龙笔",皇上穿的礼服叫作"龙袍"。许多中国古代杰出人物的传说与龙的出现有关。也许最值得一提的是,相传在孔子出生那天,他的家中请来了两条龙仪仗队为之祝贺。中国的宗教在历法中将龙的神性赋予为雨神和掌管江河湖海的神灵。在中国几乎没有几个无龙王庙的城市。
>
> ——牛顿·海耶斯《中国龙》

> Loongs with beautiful lines and symmetrical shapes can be seen everywhere in China. It is painted on silk and porcelain, woven into damask, carved on wood, embroidered on satin, cast in bronze ware, and chiseled in marble. Chinese folklore is full of countless magic legends of loongs. The emperor's throne is called "Loong Chair". The emperor used the brush is called "Loong Pen". The emperor wears a dress is called "Loong Robe". Many legends of prominent ancient Chinese figures are related to the emergence of loongs. Perhaps most note worthy, it is said that on the day of Confucius' birth, his home invited two Loong guards of honor to congratulate him. Chinese religion in the calendar endowed the divinity of the Loong to the god of rain and the god in charge of rivers, lakes and seas. There are almost few cities without the Loong King Temple in China.
>
> ——L. Newton Hayes *Loong*

> 有一句中国谚语与我们的英文谚语意思非常相近。英文中说"谈到天使你就会听到天使翅膀扇动的声音",而中文则说"说曹操曹操到"。
>
> ——贝蒂·布里奇曼《武侯祠之旅——成都南门外的一间寺庙》

> There is a Chinese proverb which is very similar in meaning to our English proverb. It is said in English, "Speak of angels and you will hear the sound of their wings." In Chinese, they say, "Speak of the devil."
>
> ——Betty Bridgman *Wu Hou Temple Trip—A Temple outside the South Gate of Chengdu*

而生活在四川的这些 CS 学校的孩子们对竹子的了解更是让我们看到了这些孩子们是怎样热爱生活、热爱四川的。他们这样赞美竹子:"四川的竹子有无穷无尽的用途。老妇用竹子的干笋壳做成鞋底,美食家用鲜竹笋烹制美味,盐业以它用作盐水管道，家具制造商则广泛用竹子制成各种实用家具，春季博览会有上百种用竹子做成的小器具，非常漂亮。竹子优雅而美丽，无论是反射在稻田里还是在月光映照下的竹影，都令人难以忘怀。中国四川农舍总是围绕着竹林，美丽和实用结合，是那个时代中国的特征。"

The CS children living in Sichuan's understanding of bamboo showed us how children love life and Sichuan. They praised bamboo like this, "Bamboo in Sichuan has endless uses. The old women use the dried bamboo shoot shell of bamboo to make shoe soles. Gourmet families cook delicious fresh bamboo shoots. The salt industry uses it as a salt water pipe. Furniture manufacturers widely use bamboo to make all kinds of practical furniture. There are hundreds of small appliances made of bamboo in the spring Expo, which are very beautiful. Bamboo is elegant and beautiful. Whether it is reflected in the rice field or in the moonlight, it is unforgettable. Sichuan farmhouses in China are always surrounded by bamboo forests. The combination of beauty and practicality is the characteristic of China at that time."

黄思礼在加拿大学校担任校长一职长达25年，他特别钟情于中国文化，并擅长绘画，他笔下的作品很多都是四川秀丽的自然风光，他的绘画才能甚至还用在了首次四川广汉（汉州）三星堆遗址考古发掘后的研究中。

Mr. Walmsley served as the Principal of the Canadian School for 25 years. He was particularly fond of Chinese culture and was good at painting. Many of his works are beautiful natural scenery in Sichuan Province. His painting skills were even used in the research after first archaeological excavation of Sanxingdui Site in Guanghan (Hanzhou), Sichuan Province.

1934年华西协合大学博物馆馆长葛维汉组织了对三星堆遗址的首次考古发掘，发掘到几百件玉器、石器和陶器，黄思礼得知后与葛维汉联手从出土的玉器、石器和陶器的颜色和纹饰来进行研究，研究结果也发表在当年葛维汉撰写的《汉州发掘初步报告》一文中。

In 1934, Mr. D. C. Graham, Curator of the Museum of West China Union University, organized the first archaeological excavation of Sanxingdui Site, and discovered hundreds of various jades, stone tools and pottery. After learning that, Mr. Walmsley and .Mr. Graham jointly studied the colors and patterns of the unearthed jades, stone tools and pottery. The results were also published in the article *PRELIMINARY REPORT OF THE HANCHOW EXCAVATION* written by Mr. Graham that year.

1948年5月，华西协合大学博物馆馆长葛维汉退休回美国，黄思礼接任馆长，同年8月他应邀回加拿大多伦多大学东亚研究系从事中国文化的教学工作。

In May, 1948, the museum curator Mr. Graham retired to the United States. Mr. Walmsley took over as the curator. In August of the same year, he was invited to return to the Department of East Asian Studies of the University of Toronto in Canada to teach Chinese culture.

1973年黄思礼完成了启真道撰写华西协合大学校史的遗愿。启真道是该校的创始人之一启尔德的儿子，1895年出生于四川乐山，在成都长大，回到加拿大接受了高等教育后，返回华西协合大学担任生理学教授，后来担任医牙学院院长。启真道退休后计划撰写华西协合大学校史，遗憾的是他还没有开始写就去世了。这件事就落在了其妹夫黄思礼的肩上，黄思礼不负众望写出了《华西协合大学》一书并出版发行，为我们留下了不可多得的史料。1999年该书由西南民族大学教授何启浩、秦和平翻译成中文在中国出版发行。

1989年，黄思礼在加拿大多伦多去世，终年92岁。

In 1973, Mr. Walmsley completed the last wish of Dr. Leslie Kilbron to write the history of West China Union University. Dr. Leslie Kilbron was the son of Dr. O. L. Kilbron, one of the founders of the University. Leslie Kilbron was born in Le Shan, Sichuan in 1895, and grew up in Chengdu. After returning to Canada for higher education, he returned to West China Union University as a professor of physiology and later as the Dean of the Medical and Dental College. After his retirement, he planned to write the history of West China Union University. Unfortunately, he died before he started writing. The writing plan fell on the shoulders of his brother-in-law Lewis C. Walmsley. Mr. Lewis C. Walmsley lived up to the expectations of the public, wrote and published *West China Union University*, which left us rare historical materials. In 1999, the book was translated into Chinese by He Qihao and Qin Heping, professors of Southwest Minzu University and published in China.

In 1989, Mr. Walmsley died in Toronto, Canada, at the age of 92.

至今留在中国的加拿大学校孩子伊莎白
Ms. Isabel —A CS Child Still Remains in China

在众多从华西坝加拿大学校毕业的孩子中，今年107岁住在北京的加拿大人伊莎白是一位特别的女士，她1933年从华西坝加拿大学校毕业后回加拿大接受高等教育后回到成都。1947年，伊莎白应中共中央外事组的邀请前往石家庄南海山外事学校（现北京外国语大学）任教并一直留在中国，是新中国英语教学的拓荒人。

Among the many children who graduated from Huaxiba Canadian School, a 107 year old Canadian, Miss Isabel Joy Brown living in Beijing, is a special lady. After graduating from Huaxiba CS in 1933, she returned to Canada for higher education then returned to Chengdu. In 1947, at the invitation of the Foreign Affairs Group of the CPC Central Committee, she went to Shijiazhuang Nanhai shan Foreign Affairs School (now Beijing Foreign Studies University) to teach and stayed in China. She is a pioneer of English Teaching in New China.

伊莎白的家就在华西坝上，她的童年和少年大部分时间都是在华西坝上度过的，每年的暑期父母都要带家人到彭州白鹿镇或峨眉山度假，因而伊莎白从小就喜欢田野生活。1933年，18岁的伊莎白从华西坝上加拿大学校完成了她的高中学业后回到加拿大，就读于多伦多大学维多利亚学院学心理学，在校期间她喜欢

成都华西坝加拿大学校师生在学校楼前合影。伊莎白（后第3排左2）1933年高中毕业于此校。柯马凯提供。

Teachers and students of Chengdu Huaxiba Canadian School took a group photo in front of the school building. Ms. Isabel (rear row 3 left 2) graduated from the high school of CS in 1933. Michael Crook Provided.

上了人类学，她还选修了人类学课程。1938年，伊莎白获得该学院儿童心理学硕士学位。伊莎白一心想做一名人类学家，她决定先回到四川做一项人类学调查，待有一些田野调查的实际经历后，再赴英国跟人类学家奠基人之一马林诺夫斯基攻读博士学位。

    Miss Isabel's home was in Huaxiba, where she spent most of her childhood and youth. Every summer, her parents took their families to Pengzhou Bailu Town or Mount Emei for vacation, so Isabel likes field life since she was a child. In 1933, 18-year-old Isabel returned to Canada after completing her high school studies from the Canadian School in Huaxiba to study psychology at Victoria College of the University of Toronto. She enjoyed anthropology while in school, and she took anthropology courses. In 1938, Miss Isabel received a master's degree in child psychology from the college. She wanted to be an anthropologist and decided to return to Sichuan to do an anthropological survey. After some fieldwork, she would go to Britain to study for her doctorate under Dr. Bronislaw Malinowski, one of the founders of anthropologists.

1939年，伊莎白回到成都并获得了教会的资助研究阿坝藏族部落，她来到理县。在近一年的田野调查活动中，伊莎白不久就初步掌握了当地少数民族的土话，能与当地人交流，还拍摄了不少反映当地民风民俗的照片。

In 1939, Miss Isabel returned to Chengdu and received funding from the church to study the Aba Tibetan tribe, and she came to Lixian County. During nearly a year of fieldwork, Isabel soon initially grasped the local minority dialect, was able to communicate with local people, and also took a lot of photos reflecting the local folk customs.

当时金陵大学、金陵女子文理学院、燕京大学和齐鲁大学在华西坝上与华西协合大学联合办学，伊莎白的妹妹在金陵大学教英文。有一天，妹妹生病了，正好伊莎白在家，她就代替妹妹去讲课，在办公室伊莎白遇到了英国人大卫·柯鲁克。柯鲁克生于英国伦敦一个犹太家庭，1934年柯鲁克毕业于哥伦比亚大学，1935年加入英国共产党，1936年前往西班牙参加国际纵队，投身到反法西斯战争中去。1937年柯鲁克读了美国记者埃德加·斯诺撰写的《西行漫记》（现译为《红星照耀中国》）一书，产生了到中国看看斯诺书中所描写的延安的想法。1938年他来到了上海，因没有路费去延安，只好在上海圣约翰大学教书，不久他应聘到成都金陵大学任教。

1939年伊莎白的父母为女儿去做田野考察送行。柯马凯提供。
In 1939, Miss Isabel's parents sent off her on a fieldwork trip. Michael Crook provided.

At that time, University of Nanking, Ginling College, Yenching University and Cheeloo University cooperated with West China Union University in Huaxiba, and Miss Isabel's sister taught English at University of Nanking. One day, her sister was ill, just when Miss Isabel was at home. She replaced her sister to lecture. In the office she met the British David Crook. Mr. Crook was born in a Jewish family in London, England. Mr. Crook graduated from Columbia University in 1934. He joined the Communist Party of Britain in 1935, and went to Spain in 1936 to join the International Brigades to devote himself to the Anti-Fascist War. In 1937, after reading the book *Red Star over China* written by American journalist Edgar Snow, Mr. Crook wanted to come to China to see Yan'an described in Snow's book. In 1938, he came to Shanghai, but had no travel fee to go to Yan'an. He had to teach at St. John's University in Shanghai. Soon, he was hired to teach at University of Nanking in Chengdu.

1940年，伊莎白与上海沪江大学社会学系毕业生俞锡玑两人到重庆璧山县兴隆乡做人类学调查，她们对1 497户家庭逐户做家访调查，经过近一年的人类学调查，完成了前期的工作。

In 1940, Miss Isabel and Miss Yu Xiji, a graduate of the Sociology Department of University of Shanghai went to Xinglong Township, Bishan County, Chongqing for anthropological investigation. They visited, 1497 families one by one. After nearly a year of anthropological investigation, they completed the preliminary work.

1942年，英、美、苏、中等结成反法西斯同盟，共产国际号召凡是盟国的共产党人都应该回国去参军，伊莎白与柯鲁克决定回英国参加反法西斯战争，而且在英国伊莎白也可以拜马林诺夫斯基为师攻读人类学博士学位，随即他们俩回到英国。他们在英国结婚后，柯鲁克到印度、缅甸等国做情报工作，而伊莎白先到一家制造子弹的军工厂工作，并且加入了英国共产党，后来她参加了加拿大女兵军团。

In 1942, Britain, the United States, the Soviet Union and China formed an Anti-Fascist Alliance. The Communist International called on all communists of the allied countries to return home to join the army. Miss Isabel and Mr. Crook decided to return to Britain to participate in the Anti-Fascist War. In Britain, Miss Isabel could also study for a doctorate in anthropology following Dr. Malinowski. Then they returned to Britain. After they got married in Britain, Mr. Crook went to India, Burma and other countries to do intelligence work. Ms. Isabel first worked in a bullet-making military factory joined the British Communist Party. Later, she joined the Canadian Women's Corps.

在英国期间，伊莎白到了伦敦经济学院才知道人类学家马林诺夫斯基已经去世，伊莎白就想跟随马林诺夫斯基的学生雷蒙德·弗思学习人类学。弗思当时还在英国海军的情报机关服役，他看了伊莎白的简历，特别是兴隆场的调查资料之后，同意在战争结束后指导她攻读人类学博士学位。

While in the UK, Ms. Isabel arrived at the London School of Economics and knew that anthropologist Dr. Malinowski had died. She wanted to study anthropology following Dr. Malinowski's student Dr. Raymond Firth. At that time, Dr. Firth was serving in the British Navy's intelligence. He looked at Ms. Isabel's resume, especially the investigation data of Xinglong Township, and agreed to guide her to a Ph.D. in anthropology after the war.

1945年5月，欧洲战事结束，伊莎白退役后获得了两年的资助进入伦敦经济学院，在弗思的指导下攻读人类学博士学位。不久柯鲁克也从英国空军退役，当他知

2019年6月，104岁高龄的伊莎白回四川寻访故地，笔者与伊莎白和她的儿子柯马凯在当年伊莎白就读的华西坝加拿大学校楼前合影。这是伊莎白从20世纪40年代初离开华西坝之后第二次回乡探访。1976年伊莎白回四川故地重游期间，她去了小时候生活的很多地方，唯独没有找到她母亲饶珍芳当年任职的弟维小学和她创办的弟幼稚园，而这所幼稚园是伊莎白曾经就读过的幼稚园。笔者受伊莎白儿子柯马凯之托经过多方查询最终找到了这所小学和幼稚园。所以，这一次的成都之旅真正圆了伊莎白的故地重访之梦。咸亚男提供。

In June, 2019, Ms. Isabel, aged 104, returned to Sichuan to visit her hometown. The author took a photo with Ms. Isabel and her son Michael Crook in front of the building of Huaxiba Canadian School where Isabel once studied. This was Ms. Isabel's second visiting hometown since she left Huaxiba in the early 1940s. When Ms. Isabel returned to her hometown of Sichuan in 1976, she went to many places where she lived as a child, but she didn't find the Dewey Primary School where her mother Ms. M. H. Brown worked and the Dewey Kindergarten her mother founded, which was the kindergarten that she once attended. Entrusted by Ms. Isabel's son Michael Crook, the author finally found the primary school and the kindergarten after many inquiries and visitings. Therefore, this trip to Chengdu truly fulfilled Ms. Isabel's dream of revisiting her hometown. Provided by Qi Yanan.

道退役军人和家属可获得旅费资助返回他们参军前的居住地方后，他与伊莎白决定回中国，到解放区去看看，写一本"关于中国农村经济和社会变革的书"，伊莎白也把此次调查作为自己的人类学博士论文项目。

In May, 1945, when the war in Europe ended, Ms. Isabel left the army and received a two-year fellowship to study for a doctorate in anthropology at the London School of Economics, under the direction of Dr. Firth. Soon Mr. Crook also left the British Air Force. When he knew that veterans and their families can get travel funds to return to where they lived before they joined the army, he and Ms. Isabel decided to return to China, to go to the liberated areas to write a book on China's economic and social changes in rural areas, and Ms. Isabel also took this investigation as her doctoral thesis project in anthropology.

1947年，伊莎白和柯鲁克带着英国共产党的介绍信来到中国，被安排到解放区河北十里店调查采访当地的土地改革。近一年的社会调查，他们收集了大量的资料，正准备回英国写书时，中共中央外事组副组长王炳南邀请他们夫妇俩留下来为新中国培养英语人才，他们欣然同意了。1948年，中共中央在刚解放的石家庄附近建立了一所外事干部培训学校，伊莎白和柯鲁克夫妇前往该校担任英语教学工作。1949年中华人民共和国成立了，外事干部培训学校迁到北京，随后改名为外国语学校（即现在的北京外国语大学）。柯鲁克担任英语部副主任和英语教研室主任，伊莎白担任二年级英语口语教师。此后伊莎白和柯鲁克就留在了中国，他们是新中国英语教学的拓荒人。

In 1947, Ms. Isabel and Mr. Crook came to China with a letter of introduction from the Communist Party of Britain, and were assigned to the liberated area of Shilidian Village, Hebei to investigate and interview the local land reform. After nearly a year of social investigation, they collected a lot of information and were just preparing to return to the UK to write a book when Mr. Wang Bingnan, Deputy Head of the Foreign Affairs Group of the CPC Central Committee, invited the couple to stay to train English talents for New China. They readily agreed. In 1948, the Central Committee of the Communist Party of China established a training school for foreign affairs cadres near the newly liberated Shijiazhuang, and Ms. Isabel and Mr. Crook went to teach English. In 1949, when the People's Republic of China was established, the Foreign Affairs Cadre Training School was moved to Beijing, and it was later renamed the Foreign Language School (now Beijing Foreign Studies University). Mr. Crook served as Deputy Director of the English Department and Director of the English Teaching and Research Section, with Ms. Isabel as a second-grade oral English teacher. Since then, Ms. Isabel and Mr. Crook have stayed in China and are pioneers of English Teaching in New China.

伊莎白和柯鲁克夫妇在从事教学的课余时间把他们在十里店收集的资料合作撰写了《十里店——中国一个村庄的革命》一书，于1959年在英国伦敦出版。

In their spare time, using the materials they had collected in Shilidian village, Ms. Isabel and Mr. Crook jointly wrote the book *Revolution in a Chinese Village: Ten Mile Inn* published in London in 1959.

伊莎白和柯鲁克一直在北京外国语大学从事英语教学。2000年，柯鲁克在北京去世，终年90岁。伊莎白始终没有忘记她想做人类学家的梦想，古稀之年仍然整理她年轻时在重庆璧山县兴隆乡作人类学调查的资料。2013年，在伊莎白98岁高龄之际，中华书局以《兴隆场——抗战时期四川农民生活调查（1940—1942）》为书名，出版了伊莎白与俞锡玑两人在70多年前做的人类学田野调查报告，为我们留下了那段历史的鲜活记忆。

Ms. Isabel and Mr. Crook have been teaching English at Beijing Foreign Studies University. Mr. Crook died in Beijing in 2000 at the age of 90. In 2013, when Ms. Isabel was 98 years old, Zhong Hua Book Company published the book *Xinglongchang——Investigation on Peasants' Life in Sichuan During the Counter-Japanese War (1940-1942)*. This anthropological fieldwork report made by Ms. Isabel and Ms. Yu Xiji over 70 years ago has left us a vivid memory of that period of history.

2019年6月，104岁的伊莎白在三个儿子的陪同下从北京来到成都华西坝寻访故地，她来到了她儿时居住的老房子校南路2号、上过学的加拿大学校……伊莎白还去了她的出生地成都四圣祠街，参观了她母亲当年出任校长的弟维小学等地方。

In June, 2019, at the age of 104, accompanied by three sons, Ms. Isabel came to Huaxiba from Beijing to visit her old site. She came to No. 2 Xiao Nanlu where she lived, Chengdu Canadian School where she studied...She also went to Sishengci North Street, Chengdu, her birthplace and visited Dewey Primary School, where her mother was the principal.

金陵女子文理学院宿舍——紫金宫
Ginling College Dormitory—Zijin Palace

1945年9月，金陵女子文理学院在紫金宫楼前草坪上举行体育运动表演赛，四周围满了观众，就连大楼阴台上也站有观众在观看。图片来源：耶鲁大学。

In September, 1945, Ginling College held a sports performance competition on the lawn in front of the CS building, surrounded by spectators, even on the shade platforms of the building were spectators. Photo Source: Yale University.

抗战时期加拿大学校的师生为了躲避日机的轰炸外迁成都郊县仁寿，CS学校大楼就借给了内迁成都华西坝的金陵女子文理学院使用，成为金陵女子文理学院的第二宿舍。

During the Counter-Japanese War, the teachers and students of CS moved to Renshou, a suburb county of Chengdu, to avoid the bombing of Japanese planes. The CS building was lent to Ginling College and became the Second Dormitory of the college.

1938年金陵女子文理学院刚来华西坝时，由华西协合大学在女生院旁提供了一块空地，教育部与四川省政府出资金为金陵女子文理学院修建了一栋两层楼的简易宿舍，也就是第一宿舍。通常新生都住第一宿舍，高年级学生才搬到第二宿舍，第二宿舍比第一宿舍的条件要好很多，因而被大家称为紫金宫。

In 1938, when Ginling College just came to Huaxiba, West China Union University provided an open space next to the Woman' College. And the Ministry of Education and the Provincial Government provided funds to build a two-story simple dormitory for Ginling College, namely the First Dormitory of Ginling College. Usually freshmen lived in the First Dormitory, and only the seniors moved to the Second Dormitory. The Second Dormitory was much better conditioned than the first dormitory, so it was called Zijin Palace.

CS学校大楼前有一块很大的草坪，自从该楼成为金陵女子文理学院的第二宿舍后，草坪就成了金陵女子文理学院学生体育活动与跳舞的场所，特别是体育和舞蹈表演比赛更是引来大量的观众观看，成为华西坝上一道亮丽的风景线。通过下面两则当年的新闻报道可以了解当时的盛况：

There was a large lawn in front of CS building. Since the building became the Second Dormitory of Ginling College, the lawn had become a place for sports activities and dance of students of Ginling College. Especially, sports and dance performance competitions attracted a large number of spectators to watch, and became a beautiful scenery in Huaxiba. The following two news reports told the grand occasion.

　　上月二十六日天空飞着微雨，气候又变得有些寒冷。坝上的人们都向着一条路上走去。因为那天是金女大的体育日。第二宿舍体育场早就挤满了人。翘着头、伸着颈，虽然有点疲乏，而每个人带回去的却是快乐的回忆。

　　金女大同学穿着运动服坐在运动场一边。司令台前陈列着几束鲜花，这是留着献给未来的姿势皇后们。

　　节目开始。每次表演之前，由体育系主任陈瑢采报告节目。清亮的声调，简单的英语，由扩音机中传来特别响亮。先由高年级的同学依次表演。体育系同学的垫上运动，精彩动人。有的从跳箱上翻过去。有的从上面跳过去，都像飞鸟似的灵敏。有的从并排六个人上翻过去。博得观众异口同声的赞美说："这真不容易！"还有"青蛙跳"，长长的一列，很有趣。

　　二年级同学的木棍操，姿势步法异常整齐，犹如公孙氏之剑。团体操八人一组轮流舞着。有长长的黑辫也随她的步法在空中左右回旋。

　　——《燕京新闻（成都版）》1944年4月1日第10卷第21期第4版

On June 26, the weather became chilly with light rain. All the people in the Huaxiba turned to a road, for it was sports day of Ginling College. The playground of the Second Dormitory was already full of people, their heads up, necks out, although a little tired, but everyone took home happy memories.

Students dressed in sportswear sitting on the side of the playground. In front of the judges' stand, there were several bundles of flowers, which were reserved for the future posture queens.

At the beginning of the program, before each performance, Chen Liancai, head of the Department of Physical Education, reported the program. The clear tone and simple English were particularly loud from the loudspeaker. First, senior students performed in turn, and the mat sports of students in the Department of Physical Education were wonderful and moving. Some turned over from the jump box, and some jumped over from the jump box top, all as sensitive as birds. Some turned over from six side by side people, which won the praise of the audience, "It's really not easy!" There was the frog jump, a long long line, very interesting.

The second grade students' wooden stick gymnastics posture and footwork were very neat. Like Gongsun's sword. A group of 8 people danced in turn, with long black braids whirling left and right in the air with the dancer's steps.

——*Yenching News (Chengdu Edition)*, April 1, 1944, Volume 10, Issue 21, Edition 4

第二则报道的是金陵女子文理学院在1945年5月19日下午举行的五月花柱节活动，活动在第二宿舍前大草坪举办。金陵女子文理学院通常每年的五月花柱节是对外开放的，大家都可以来观看，不售票。但是这一次，她们专门印制了观看券出售，把售票收入用来给前线战士制作草鞋以及慰问出征的抗日士兵。

The second news reported the Maypole Festival held by Ginling College on the afternoon of May 19, 1945, which was held on the big lawn in the front off the Second Dormitory. Ginling College annual Maypole Festival was usually open to the public, and everyone could come to watch and sold no tickets. But this time, they specially printed tickets to sell, and used the ticket proceeds to make straw sandals for the front-line soldiers and comfort the counter-Japanese soldiers.

在表演日，夕阳尚未下山，第二宿舍门前早已人山人海，这些人一方面是来欣赏舞蹈的，另一方面也是爱国的表现，200元一张的票价大家简直不当一回事，而且还有花500元从黑市买票的呢!

许多行政长官也久慕金陵女子文理学院的盛名，纷纷进场。一个方形的广场，一架钢琴，一架扩音器，体育界的名人坐在旁边担任裁判。四点钟的时候，表演开始了，在绿草如茵的地上，映着夕阳的斜晖，男女（男的是由女的假扮的）翩翩起舞，那一种美是难以用言语来形容的，四周的观众，寂然无声，接踵而至，听到的是琴声、播音员的报告和开来拉（cello，大提琴）的声音。各班级的土风舞比赛，由每班24人共同表演，动作的整齐，姿势的优美，表情的逼真，加之服装的形形色色，花的被面用来做裙子，嵌着各种颜色的纸花边；漂亮的领结嵌在领子上，这一点你会看不出它是纸做的。在这个物力匮乏的年代，你不得不佩服她们的设计天才……

各班级比赛完后，接着便是正式的舞蹈表演，名称是《舞蹈的演进》，

华西坝老建筑的前世今生（汉英对照）
The Stories of the Historic Architecture of Huaxiba (Chinese-English Bilingual Edition)

这张从紫金宫楼上拍摄的金女大体育运动表演赛的照片，女生们整齐划一的运动展示了她们的团队精神。图片来源：耶鲁大学。

This photo of the Ginling College sports performance competition taken from the upstairs of the CS building shows that the girls' neat and uniform sports show their team spirit. Photo Source: Yale University.

270

拾柒 中加两国人民友谊的见证——志德堂

由穿着兽皮、披着树叶的原始舞蹈起，一步一步地演进，一直到现代舞，你可以看到原始舞是简单的，不过随着跳舞的演进，慢慢地舞蹈动作愈来愈复杂，姿态愈来愈美妙，服装愈来愈讲究。象征着时代的演进，其中尤以埃及舞、吉普赛舞、西班牙舞，惟妙惟肖，情景逼真，俨然如跃身其中。

最后为场面最大的竿舞，每级代表16人围成圆形，四年级居中，环绕竹竿，竿上缠着代表学校的紫白绳带各16条，紫白相间，飞舞空中，略成伞形，数十人绕竿而舞，亭亭玉立，别具风格，观众眼花缭乱，忘其所以。

——1945年第19期《中华全国体育协进会体育通讯》

On the performance day, the setting sun had not yet gone down and there were already a large number of people in front of the Second Dormitory. These people on the one hand were to enjoy the dance, on the other hand, were also a manifestation of patriotism. The ticket price of 200 yuan was simply not a matter, and there were people who spend 500 yuan to buy tickets from the black market.

Many chief executives of the city had long admired the reputation of Ginling College and had entered the field one after another. A square square, a piano and a loudspeaker, and sports celebrities sitting next to serve as a referee. At four o'clock, the performance began. On the green grass, reflecting the setting sun, men and women (men were disguised by women) were dancing. That kind of beauty was difficult to describe in words. The audience came one after another. They were silent, watching, hearing the sound of the piano, the announcer's report and the sound of cello. The folk dance competition of each class was performed by 24 people per class. The movement was neat. The posture was beautiful. The expression was lifelike. They were wearing various clothes, with colorful quilt covers made skirts, embedded in various colors of paper lace, and beautiful bow tie was embedded in the collar, which you can not see that it is made of paper. In this age of material scarcity, you have to admire their design genius.

After the class competition, followed by the formal dance performance, the name was *The Evolution of Dance*. Starting from a primitive dance dressed in animal skins and covered in leaves, step by step evolution, to modern dance, you can see the primitive dance is simple. But with the evolution of dance, slowly, the dance actions are more and more complex, the postures are more and more wonderful, the clothes are more and more exquisite, symbolized the evolution of the time. Especially Egyptian dance, Gypsy dance, Spanish dance, vivid, lifelike scene, the audience felt dancing among them.

Finally, there was the biggest pole dance, with 16 people of grade representatives in a circle, the senior grade in the middle, surrounding a bamboo pole. There were 16 purple and white ropes representing the school wrapped around the pole. Purple and white were flying in the air, slightly into an umbrella shape. Dozens of people surrounded the pole and danced So graceful, unique style, that the audience dazzled, forgot oneself.

—*Sports Newsletter of All-China Sports Association*, No. 19, 1945

中国公共卫生之父——陈志潜
Dr. C. C. Chen, the Father of Public Health in China

陈志潜1903年9月出生于成都，父亲陈可大是清末秀才，母亲在他4岁时就因病去世了，家里的其他几位亲人在他小时候也相继过世了，这给了他刻骨铭心的痛，他立志将来要学医当医生。1921年陈志潜考入了北平协和医学院，毕业后留校成为一名内科医生。

C. C. Chen was born in Chengdu in September of 1903. His father Chen Keda was a scholar in the Qing Dynasty. His mother died of illness when he was 4 years old, and several relatives in his family died one after another when he was a child, which gave him unforgettable pain. He was determined to study medicine and become a doctor in the future. In 1921, C. C. Chen was admitted to Peking Union Medical College. After graduation, he stayed in the college to become a qualified physician.

陈志潜还在北平协和医学院读书期间就关注到普通民众的健康问题，他开始在导师兰安生的指导下学习公共卫生学。兰安生是加拿大人，世界著名公共卫生学家，1809年出生在浙江一位医学传教士家庭。1917年兰安生毕业于美国密歇根大学医学院，之后他子承父业来到中国从事公共卫生事业。1921年兰安生在北平协和医学院创建了中国最早的公共卫生学系。陈志潜逐渐认识到国家的复兴与公共卫生的关系极为密切，预防工作更为重要，他放弃了当内科医生可以获得较高收入的机会，决心不计报酬而献身公共卫生事业。后来陈志潜回忆道："选择公共卫生作为终身职业需要强烈的社会责任感和决心，若没有爱国的热情和远大的理想为动力，选择这个专业是毫无意义的。这项工作要求牺牲研究工作在学术上的成就和私人开业在经济上的收入，而且要在很困难的条件下开展工作，而不可能在较为舒适的城市里。"

2004年在中国公共卫生之父陈志潜教授百年诞辰之际，四川大学在志德堂旁新建的华西公共卫生学院大楼门厅前竖立了一尊陈志潜铜像。咸亚男摄于2022年。
In 2004, on the occasion of the 100th Anniversary of the Birth of Professor Chen Zhiqian (C.C.Chen), the father of Public health in China, Sichuan University built a bronze statue of Chen Zhiqian in front of the lobby of the new West China School of Public Health building next to CS building. Photo by Qi Yanan in 2022.

Dr. C. C. Chen also focused on the health problems of ordinary people during his studies at the Peking Union Medical College. He began to study public health under the guidance of his mentor John B. Grant. Dr. Grant is a Canadian and world-famous public health scientist. He was born in 1809 to a medical missionary family in Zhejiang Province, China in 1809. He graduated from the University of Michigan School of Medicine in 1917. After that, he followed his father's footsteps and came to China to work in public health. In 1921, Dr. Grant founded the earliest Department of Public Health in China at Peking Union Medical College. Dr. C. C. Chen gradually realized that the revival of a country is very closely related to public health and that prevention is more important. He gave up the higher salary he could get as a physician and decided to devote himself to health care regardless of pay. Later, Dr. C. C. Chen recalled:"Choosing public health as a lifelong career requires a strong sense of social responsibility and determination. Without patriotic enthusiasm and lofty ideals, it is meaningless to choose this major. This work requires the sacrifice of academic research achievements and financial income from private practice, and the need of work under difficult conditions, not possible in more comfortable cities."

毕业后，兰安生安排陈志潜到美国哈佛大学公共卫生学院学习。1931年陈志潜回国后，兰安生又把他推荐给我国平民教育家和乡村建设家晏阳初，晏阳初聘请陈志潜到河北定县平民教育促进会的农村建设实验区担任卫生教育部主任，在这个职位上，陈志潜做了10年的农村卫生工作，领导了史无前例的定县卫生保健系统的设计与实施。陈志潜在定县创立的农村三级保健网，开展保健服务和健康教育，成为具有世界影响的"定县模式"，对我国公共卫生事业的发展起到了奠基作用。1958年我国政府采纳了定县的经验，建立了全国性的乡村卫生保健系统，由此产生的"赤脚医生"为农村广大农民提供了最基本的卫生保健，这些在中国取得的经验也为国际公共卫生学界所推崇。

After graduation, Dr. Grant arranged for Dr. C. C. Chen to study at the School of Public Health of Harvard University. After he returned to China in 1931, Dr. Grant recommended him to Dr. Y. C. James Yen, the civilian educators and rural construction expert of China. Dr. Y. C. James Yen hired Dr. C. C. Chen to the Rural Construction Experimental Area of Hebei Dingxian County Civilian Education Promotion Association served as the Director of the Section of Health Education. In this position, he did 10 years of rural health work, led the unprecedented design and implementation of the Dingxian County health care system. The three-level rural health care network established by Dr. C. C. Chen, which had carried out health services and health education, had become a "Dingxian Model" with world influence and played a foundation role in the development of China's public health undertakings in China. In 1958, the Chinese Government adopted the experience of Dingxian and established a national rural health care system. The resulting barefoot doctors provided the most basic health care for rural farmers, and these experience in China is also highly respected by the international public health community.

"卢沟桥事变"后，陈志潜在定县搞的农村卫生工作被迫停止了。当时陈志潜担任北平第一卫生所和定县农村卫生教育部的负责人，为了躲避日本人，他秘密离开北平南下从事抗战救护工作。

After the Lugouqiao Incident, Dr. C. C. Chen's rural health work in Dingxian county was forced to stop. At that time, Dr. C. C. Chen was the head of the First Health Center in Peking and the Rural Health and Education Department of Dingxian County. In order to avoid the Japanese, he secretly left Peking and went south to engage in the rescue work of the Counter-Japanese War.

1939年，陈志潜应四川省政府的邀请回到成都组建四川省卫生实验处并出任处长，同时他还兼任中央大学医学院、华西协合大学医学院教授。陈志潜回忆那时所做的工作有："与医疗援救工作有关的公共卫生服务的框架已经形成。伤员的临时医院、隔离医院、各种妇幼保健诊所，一处护士、助产士和公共卫生人员的培训中心，以及农村教学机构都已经建立起来。此外，还有流动的防疫医疗队。"

In 1939, at the invitation of the Sichuan Provincial Government, Dr. C. C. Chen returned to Chengdu to establish the Sichuan Provincial Health Experimental Department and served as the Director. Meanwhile, he was also a professor at the School of Medicine of West China Union University and Central University School of Medicine. Dr. C. C. Chen recalled that the work done at that time was that "the framework for public health services related to our medical rescue work has been formed. Temporary hospitals for the wounded, isolation hospitals, various maternal and child health clinics, a training center for nurses, midwives and public health personnel, and rural teaching institutions have all been established. In addition, there are mobile epidemic prevention and prevention medical teams."

抗战胜利后，陈志潜被调到重庆筹建国立重庆大学医学院并出任院长一职，这是四川第一所政府办的医学院。

After the victory of the Counter-Japanese War, Dr. C. C. Chen was transferred to Chongqing to become the Dean of the National Chongqing University Medical School, the first government-run medical school in Sichuan.

1952年全国高等院校调整，重庆大学医学院在调整中被撤销，并入四川医学院，陈志潜又回到成都，在四川医学院医学系担任卫生学教授。1955年四川医学院建立了卫生系，陈志潜被任命为代理系主任进入第七教学楼办公。陈志潜在第七教学楼工作了45年，直到2000年去世，被人们誉为"中国公共卫生之父"。

In 1952, during the adjustment of the departments of colleges and universities nationwide, Chongqing University Medical School was removed and merged into Sichuan Medical College. Dr. C. C. Chen returned to Chengdu to serve as a professor of hygienic in the Department of Medicine of Sichuan Medical College. In 1955, Sichuan Medical College established the Department of Health. Dr. C. C. Chen was appointed as the Acting Dean to work in the Seventh Teaching Building. He worked in the Seventh Teaching Building for 45 years until his death in 2000, and was known as the father of public health in China.

陈志潜的导师兰安生的儿子、联合国儿童基金会前主席格兰特这样评价陈志潜："陈志潜教授致力于卫生工作50多年，对世界卫生工作做出了不可估量的贡献。这些贡献至今仍在促进着中国人民的健康和幸福，同样也在相当程度地改善着世界其他发展中国家人民的健康和幸福。"他还这样说："由于你在30年代时首创的初级卫生保健的基本概念，使今天千百万人得以生存。使全民受益于有组织的卫生服务，而不是只为少数人服务的革命思想是您在本世纪在此星球上的重要贡献。家父兰安生博士和我以能与您在一起工作而感庆幸。"

Mr. James P. Grant, son of C.C. Chen's mentor Grant and former head of the United Nations Children's Fund, said of Dr. Chen, "Professor Chen has been committed to the health work for more than 50 years, and has made an immeasurable contribution to the world health work. These contributions are still promoting the health and well-being of the Chinese people, and are also fairly improving the health and happiness of the people of other developing countries in the world." He also said, "The basic concept of primary health care, which you pioneered in the 1930s, has enabled millions of people to survive today. The revolutionary idea of making the whole people benefit from organized health services rather than serving only a few people is your important contribution to this planet in this century. My father, Dr. John B. Grant, and I are glad to work with you."